7/84

THE CONTEST
PROBLEM BOOK III

Annual High School Contests
1966–1972
of
The Mathematical Association of America
Society of Actuaries
Mu Alpha Theta
National Council of Teachers of Mathematics
Casualty Actuarial Society

NEW MATHEMATICAL LIBRARY

PUBLISHED BY

THE MATHEMATICAL ASSOCIATION OF AMERICA

The New Mathematical Library (NML) was begun in 1961 by the School Mathematics Study Group to make available to high school students short expository books on various topics not usually covered in the high school syllabus. In a decade the NML matured into a steadily growing series of some twenty titles of interest not only to the originally intended audience, but to college students and teachers at all levels. Previously published by Random House and L. W. Singer, the NML became a publication series of the Mathematical Association of America (MAA) in 1975. Under the auspices of the MAA the NML will continue to grow and will remain dedicated to its original and expanded purposes.

THE CONTEST PROBLEM BOOK III

Annual High School Contests

1966–1972

compiled and with solutions by

Charles T. Salkind

Polytechnic Institute of Brooklyn

and

James M. Earl

University of Nebraska at Omaha

25

THE MATHEMATICAL ASSOCIATION

OF AMERICA

Illustrated by George H. Buehler

Fourth Printing

Library of Congress Catalog Card Number: 66-15479

Complete Set ISBN-0-88385-600-X

Vol. 25 0-88385-625-5

Manufactured in the United States of America

Contents

NEW MATHEMATICAL LIBRARY

Other titles in preparation

Preface

Problem solving is at the heart of learning mathematics; a student's ability to perceive, master and work with mathematical fundamentals is greatly enhanced by encouraging him to solve carefully designed problems. A good problem, like an acorn, contains the potential for grand development. The Committee on High School Contests, in this spirit, seeks to extend and supplement regular school work through the Annual High School Mathematics Examination. First organized in 1950 and restricted to Metropolitan New York, these examinations were sponsored nationally in 1957 by the Mathematical Association of America and the Society of Actuaries, and later cosponsored by Mu Alpha Theta (1965), the National Council of Teachers of Mathematics (1967), and the Casualty Actuarial Society (1971).

An important difference between these and some similarly motivated European competitions is that our Annual Examination aims to discriminate on *several levels,* and is not *exclusively* directed to high-ability students.

All students are welcome to participate as individuals or in teams of three from the same school. The top scorer receives the Association's Mathematics Pin award; there are also lesser awards including a Certificate of Merit for teams scoring in the upper decile regionally.

The number of contestants has grown from approximately 150,000 in 1960 to more than 350,000 in 1972 within the ten Canadian and U.S. regions; in addition there were thousands of participants abroad. One hundred of the top students took part, on May 9, 1972, in a very successful first USA Mathematical Olympiad, a 5 question, 3 hour subjective test.

The questions on each examination are grouped according to difficulty and complexity as estimated by the members of the Committee. Questions in the first two parts are meant to deal with basic concepts, those in the later parts are meant to test the application of skills to various new situations.† On the 1966 and 1967 examinations, Parts 1, 2, and 3 consisted

† For a subset of the contestants, the number of correct responses has been tabulated, and the categorization of questions according to difficulty seems justified.

1

of 20, 10, and 10 questions weighted 3, 4, and 5. On the 1968–1972 examinations, the questions are divided into four parts of 10, 10, 10, and 5 questions, respectively, with weights 3, 4, 5, and 6.†

The solutions presented here are by no means the only ones possible, nor are they necessarily superior to other alternatives. Since no mathematics beyond intermediate algebra is required, an elementary procedure is always given, even where "high-powered" alternatives are added.

The MAA Committee invites your comments.

Charles T. Salkind

James M. Earl

---

† Examinations are scored by the formula $R - \frac{1}{4}W$, where R and W denote weighted counts of correct and incorrect responses, respectively.

Editors' Preface

The editors of the New Mathematical Library, in wishing to encourage significant problem-solving at the high school level, have published the following problem collections so far: NML 5 and 17 containing all annual contest problems proposed by the Mathematical Association of America through 1965; and NML 11 and 12 containing translations of all Eötvös Competition problems through 1928 and their solutions. The present volume is a sequel to NML 17 published at the request of the many readers who enjoyed the previous MAA problem books.

The MAA contests now contain 35 problems based entirely on the standard high school curriculum. To expedite grading of the approximately 400,000 papers written, each question is worded so that exactly one of five choices offered serves as a correct answer.

Each Eötvös contest, on the other hand, contains only three problems, based on the Hungarian high school curriculum, and often requiring ingenuity and rather deep investigations for their solution.

The MAA is concerned primarily with mathematics on the undergraduate level. It is one of three major mathematical organizations in America (the other two being the American Mathematical Society, chiefly concerned with mathematical research, and the National Council of Teachers of Mathematics, concerned with the content and pedagogy of elementary and secondary mathematics). The MAA also conducts the annual Putnam Competition for undergraduate students. Its journal, The American Mathematical Monthly, is famous for its elementary and advanced problem sections.

The editors of the New Mathematical Library are glad to cooperate with the MAA in publishing this collection. They wish to acknowledge, in particular, the essential contributions of the two men who compiled and wrote solutions for the problems in the present collection: Prof. Charles T. Salkind, responsible for the contests up to the year of his

3

death, and Professor James M. Earl, who succeeded him in 1968. During his illness Professor Earl not only supervised the editing of the present volume, but, with stoic dedication, submitted the problems for the 1973 Annual Contest shortly before his death on November 26, 1972.

A few minor changes in the statements of contest problems have been made in this collection for the sake of greater clarity.

Note: The student is told to avoid random guessing, since there is a penalty for wrong answers. However, if he can definitely eliminate some of the choices, a random guess among the remaining choices will result in a positive expected score. A few examples of such elimination are indicated in the remarks appended to some solutions.

Basil Gordon

Anneli Lax

1973

Suggestions for Using this Book

This problem collection is designed to be used by mathematics clubs, high school teachers, students, and other interested individuals. Clearly, no one would profit from doing *all* the problems, but he *would* benefit from those that present a challenge to him. The reader might try himself on a whole test or on part of a test, with (or preferably without) time limitations.

He should try to get as far as possible with the solution to a problem. If he is really stuck, he should look up the answer in the key (p. 55) and try to work backwards; if this fails, the section of complete solutions should be consulted.

In studying solutions, even the successful problem solver may find sidelights he had overlooked; he may find a more "elegant" solution, or a way of solving the problem which may lead him deeper into mathematics. He may find it interesting to change items in the hypothesis and to see how this affects the solution, or to invent his own problems.

If a reader is interested in a special type of problem, he should consult the classified index.

The following familiar symbols appear in this book:

Symbol	Meaning
$\sim$	similar (if used in connection with plane figures)
$\sim$	approximately equal (if used in connection with numbers)
$\therefore$	therefore
$\equiv$	identically equal to
$<$	less than
$\leq$	less than or equal to
$>$	greater than
$\geq$	greater than or equal to
$\lvert k \rvert$	absolute value of the number k
$\triangle$	triangle
110_2	the number $1 \cdot 2^2 + 1 \cdot 2^1 + 0 \cdot 2^0$, i.e., the number 6 when written in a numeration system with base 2 instead of 10.
$\cong$	congruent
$\neq$	different from
$\perp$	perpendicular to
XY	length of the line segment XY, often denoted by $\overline{XY}$ in other books
$f(x)$	function of the variable x
$\parallel$	parallel to
$\overset{\frown}{AB}$	circular arc with endpoints A and B.

5

Problems

1966 Examination

Part 1

1. Given that the ratio of $3x - 4$ to $y + 15$ is constant, and $y = 3$ when $x = 2$, then, when $y = 12$, x equals:

   (A) $\frac{1}{8}$ (B) $\frac{3}{7}$ (C) $\frac{7}{3}$ (D) $\frac{7}{2}$ (E) 8

2. When the base of a triangle is increased 10% and the altitude to this base is decreased 10%, the change in area is

   (A) 1% increase   (B) $\frac{1}{2}$% increase   (C) 0%
   (D) $\frac{1}{2}$% decrease   (E) 1% decrease

3. If the arithmetic mean of two numbers is 6 and their geometric mean is 10, then an equation with the given two numbers as roots is:

   (A) $x^2 + 12x + 100 = 0$ (B) $x^2 + 6x + 100 = 0$
   (C) $x^2 - 12x - 10 = 0$ (D) $x^2 - 12x + 100 = 0$
   (E) $x^2 - 6x + 100 = 0$

4. Circle I is circumscribed about a given square and circle II is inscribed in the given square. If r is the ratio of the area of circle I to that of circle II, then r equals:

   (A) $\sqrt{2}$ (B) 2 (C) $\sqrt{3}$ (D) $2\sqrt{2}$ (E) $2\sqrt{3}$

5. The number of values of x satisfying the equation
$$\frac{2x^2 - 10x}{x^2 - 5x} = x - 3$$
is:

(A) zero   (B) one   (C) two   (D) three
(E) an integer greater than 3

6. AB is a diameter of a circle centered at O. C is a point on the circle such that angle BOC is 60°. If the diameter of the circle is 5 inches, the length of chord AC, expressed in inches, is:

(A) 3   (B) $\dfrac{5\sqrt{2}}{2}$ (C) $\dfrac{5\sqrt{3}}{2}$ (D) $3\sqrt{3}$ (E) none of these

7. Let $\dfrac{35x - 29}{x^2 - 3x + 2} = \dfrac{N_1}{x - 1} + \dfrac{N_2}{x - 2}$ be an identity in x. The numerical value of $N_1 N_2$ is:

(A) -246 (B) -210 (C) -29 (D) 210   (E) 246

8. The length of the common chord of two intersecting circles is 16 feet. If the radii are 10 feet and 17 feet, a possible value for the distance between the centers of the circles, expressed in feet, is:

(A) 27   (B) 21   (C) $\sqrt{389}$ (D) 15   (E) undetermined

9. If $x = (\log_8 2)^{(\log_2 8)}$, then $\log_3 x$ equals:

(A) -3 (B) $-\frac{1}{3}$ (C) $\frac{1}{3}$ (D) 3   (E) 9

10. If the sum of two numbers is 1 and their product is 1, then the sum of their cubes is:

(A) 2   (B) $-2 - \dfrac{3\sqrt{3}i}{4}$ (C) 0   (D) $-\dfrac{3\sqrt{3}i}{4}$ (E) -2

[Here i denotes $\sqrt{-1}$.]

11. The sides of triangle BAC are in the ratio 2:3:4. BD is the angle-bisector drawn to the shortest side AC, dividing it into segments AD and CD. If the length of AC is 10, then the length of the longer segment of AC is:

(A) $3\frac{1}{2}$ (B) 5   (C) $5\frac{5}{7}$ (D) 6   (E) $7\frac{1}{2}$

12. The number of real values of x that satisfy the equation

$$(2^{6x+3})(4^{3x+6}) = 8^{4x+5}$$

is:

(A) zero   (B) one   (C) two   (D) three   (E) greater than 3

13. The number of points with positive rational coordinates selected from the set of points in the xy-plane such that $x + y \le 5$, is:

(A) 9   (B) 10   (C) 14   (D) 15   (E) infinite

14. The length of rectangle $ABCD$ is 5 inches and its width is 3 inches. Diagonal AC is divided into three equal segments by points E and F. The area of triangle BEF, expressed in square inches, is:

(A) $\frac{3}{2}$ (B) $\frac{5}{3}$ (C) $\frac{5}{2}$ (D) $\frac{1}{3}\sqrt{34}$ (E) $\frac{1}{3}\sqrt{68}$

15. If $x - y > x$ and $x + y < y$, then

(A) $y < x$ (B) $x < y$ (C) $x < y < 0$ (D) $x < 0, \ y < 0$
(E) $x < 0, \ y > 0$

16. If $\dfrac{4^x}{2^{x+y}} = 8$ and $\dfrac{9^{x+y}}{3^{5y}} = 243$, x and y real numbers, then xy

equals:

(A) $\frac{12}{5}$ (B) 4   (C) 6   (D) 12   (E) -4

17. The number of distinct points common to the curves $x^2 + 4y^2 = 1$ and $4x^2 + y^2 = 4$ is:

(A) 0   (B) 1   (C) 2   (D) 3   (E) 4

18. In a given arithmetic sequence the first term is 2, the last term is 29, and the sum of all the terms is 155. The common difference is:

(A) 3   (B) 2   (C) $\frac{27}{19}$ (D) $\frac{13}{9}$ (E) $\frac{23}{38}$

19. Let s_1 be the sum of the first n terms of the arithmetic sequence 8, 12, $\cdots$ and let s_2 be the sum of the first n terms of the arithmetic sequence 17, 19, $\cdots$. Assume $n \ne 0$. Then $s_1 = s_2$ for:

(A) no value of n (B) one value of n (C) two values of n
(D) four values of n (E) more than four values of n

20.† The negation of the proposition "For all pairs of real numbers a, b, if $a = 0$, then $ab = 0$" is: There are real numbers a, b such that

(A) $a \neq 0$ and $ab \neq 0$ (B) $a \neq 0$ and $ab = 0$
(C) $a = 0$ and $ab \neq 0$ (D) $ab \neq 0$ and $a \neq 0$
(E) $ab = 0$ and $a \neq 0$

Part 2

21. An "n-pointed star" is formed as follows: the sides of a convex polygon are numbered consecutively $1, 2, \cdots, k, \cdots, n$, $n \geq 5$; for all n values of k, sides k and $k + 2$ are non-parallel, sides $n + 1$ and $n + 2$ being respectively identical with sides 1 and 2; prolong the n pairs of sides numbered k and $k + 2$ until they meet. (A figure is shown for the case $n = 5$).

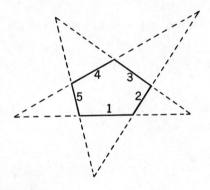

Let S be the degree-sum of the interior angles at the n points of the star; then S equals:

(A) 180   (B) 360   (C) $180(n + 2)$ (D) $180(n - 2)$
(E) $180(n - 4)$

22. Consider the statements: (I) $\sqrt{a^2 + b^2} = 0$ (II) $\sqrt{a^2 + b^2} = ab$ (III) $\sqrt{a^2 + b^2} = a + b$ (IV) $\sqrt{a^2 + b^2} = a \cdot b$, where we allow a and b to be real or complex numbers. Those statements for which there exist solutions other than $a = 0$ and $b = 0$, are:

(A) (I), (II), (III), (IV)   (B) (II), (III), (IV) only
(C) (I), (III), (IV) only   (D) (III), (IV) only
(E) (I) only

† This problem differs from the one given on the original 1966 Examination.

23. If x is real and $4y^2 + 4xy + x + 6 = 0$, then the complete set of values of x for which y is real, is:

(A) $x \leq -2$ or $x \geq 3$ (B) $x \leq 2$ or $x \geq 3$
(C) $x \leq -3$ or $x \geq 2$ (D) $-3 \leq x \leq 2$
(E) $-2 \leq x \leq 3$

24. If $\log_M N = \log_N M$, $M \neq N$, $MN > 0$, $M \neq 1$, $N \neq 1$, then MN equals:

(A) $\frac{1}{2}$ (B) 1   (C) 2   (D) 10
(E) a number greater than 2 and less than 10

25. If $F(n+1) = \dfrac{2F(n)+1}{2}$ for $n = 1, 2, \cdots$, and $F(1) = 2$,

then $F(101)$ equals:

(A) 49   (B) 50   (C) 51   (D) 52   (E) 53

26. Let m be a positive integer and let the lines $13x + 11y = 700$ and $y = mx - 1$ intersect in a point whose coordinates are integers. Then m can be:

(A) 4 only   (B) 5 only   (C) 6 only   (D) 7 only
(E) one of the integers 4, 5, 6, 7 and one other positive integer

27. At his usual rate a man rows 15 miles downstream in five hours less time than it takes him to return. If he doubles his usual rate, the time downstream is only one hour less than the time upstream. In miles per hour, the rate of the stream's current is:

(A) 2   (B) $\frac{5}{2}$ (C) 3   (D) $\frac{7}{2}$ (E) 4

28. Five points O, A, B, C, D are taken in order on a straight line with distances $OA = a$, $OB = b$, $OC = c$, and $OD = d$. P is a point on the line between B and C and such that $AP:PD = BP:PC$. Then OP equals:

(A) $\dfrac{b^2 - bc}{a - b + c - d}$ (B) $\dfrac{ac - bd}{a - b + c - d}$ (C) $-\dfrac{bd + ac}{a - b + c - d}$

(D) $\dfrac{bc + ad}{a + b + c + d}$ (E) $\dfrac{ac - bd}{a + b + c + d}$

29. The number of positive integers less than 1000 divisible by neither 5 nor 7 is:

    (A) 688   (B) 686   (C) 684   (D) 658   (E) 630

30. If three of the roots of $x^4 + ax^2 + bx + c = 0$ are 1, 2, and 3, then the value of $a + c$ is:

    (A) 35   (B) 24   (C) −12   (D) −61   (E) −63

Part 3

31. Triangle ABC is inscribed in a circle with center O'. A circle with center O is inscribed in triangle ABC. AO is drawn, and extended to intersect the larger circle in D. Then we must have:

    (A) $CD = BD = O'D$ (B) $AO = CO = OD$
    (C) $CD = CO = BD$ (D) $CD = OD = BD$
    (E) $O'B = O'C = OD$

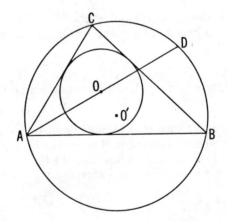

32. Let M be the midpoint of side AB of triangle ABC. Let P be a point on AB between A and M, and let MD be drawn parallel to PC and intersecting BC at D. If the ratio of the area of triangle BPD to that of triangle ABC is denoted by r, then

    (A) $\frac{1}{2} < r < 1$ depending upon the position of P
    (B) $r = \frac{1}{2}$ independent of the position of P
    (C) $\frac{1}{2} \le r < 1$ depending upon the position of P
    (D) $\frac{1}{3} < r < \frac{2}{3}$ depending upon the position of P
    (E) $r = \frac{1}{3}$ independent of the position of P

33. If $ab \neq 0$ and $|a| \neq |b|$ the number of distinct values of x satisfying the equation

$$\frac{x-a}{b} + \frac{x-b}{a} = \frac{b}{x-a} + \frac{a}{x-b},$$

is:

(A) zero   (B) one   (C) two   (D) three   (E) four

34. Let r be the speed in miles per hour at which a wheel, 11 feet in circumference, travels. If the time for a complete rotation of the wheel is shortened by $\frac{1}{4}$ of a second, the speed r is increased by 5 miles per hour. Then r is:

(A) 9   (B) 10   (C) $10\frac{1}{2}$ (D) 11   (E) 12

35.† Let O be an interior point of triangle ABC, and let $s_1 = OA + OB + OC$. If $s_2 = AB + BC + CA$, then

(A) for every triangle $s_2 > 2s_1$, $s_1 \leq s_2$
(B) for every triangle $s_2 \geq 2s_1$, $s_1 < s_2$
(C) for every triangle $s_1 > \frac{1}{2}s_2$, $s_1 < s_2$
(D) for every triangle $s_2 \geq 2s_1$, $s_1 \leq s_2$
(E) neither (A) nor (B) nor (C) nor (D) applies to every triangle

36. Let $(1 + x + x^2)^n = a_0 + a_1x + a_2x^2 + \cdots + a_{2n}x^{2n}$ be an identity in x. If we let $s = a_0 + a_2 + a_4 + \cdots + a_{2n}$, then s equals:

(A) 2^n (B) $2^n + 1$ (C) $\dfrac{3^n - 1}{2}$ (D) $\dfrac{3^n}{2}$ (E) $\dfrac{3^n + 1}{2}$

37. Three men, Alpha, Beta, and Gamma, working together, do a job in 6 hours less time than Alpha alone, in 1 hour less time than Beta alone, and in one-half the time needed by Gamma when working alone. Let h be the number of hours needed by Alpha and Beta, working together, to do the job. Then h equals:

(A) $\frac{5}{2}$ (B) $\frac{3}{2}$ (C) $\frac{4}{3}$ (D) $\frac{5}{4}$ (E) $\frac{3}{4}$

38. In triangle ABC the medians AM and CN to sides BC and AB, respectively, intersect in point O. P is the midpoint of side AC, and MP intersects CN in Q. If the area of triangle OMQ is n, then the area of triangle ABC is:

(A) $16n$ (B) $18n$ (C) $21n$ (D) $24n$ (E) $27n$

† The five choices have been slightly altered by the editors.

39. In base R_1 the expanded fraction F_1 becomes .373737···, and the expanded fraction F_2 becomes .737373···. In base R_2 fraction F_1, when expanded, becomes .252525···, while fraction F_2 becomes .525252···. The sum of R_1 and R_2, each written in the base ten, is:

(A) 24   (B) 22   (C) 21   (D) 20   (E) 19

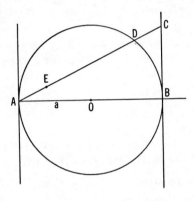

40. In this figure AB is a diameter of a circle, centered at O, with radius a. A chord AD is drawn and extended to meet the tangent to the circle at B in point C. Point E is taken on AC so that $AE = DC$. Denoting the distances of E from the tangent through A and from the diameter AB by x and y, respectively, we can deduce the relation:†

(A) $y^2 = \dfrac{x^3}{2a - x}$ (B) $y^2 = \dfrac{x^3}{2a + x}$ (C) $y^4 = \dfrac{x^2}{2a - x}$

(D) $x^2 = \dfrac{y^2}{2a - x}$ (E) $x^2 = \dfrac{y^2}{2a + x}$

† On the 1966 Examination the last sentence in the statement of Problem 40 was somewhat different.

1967 Examination

Part 1

1. The three-digit number $2a3$ is added to the number 326 to give the three-digit number $5b9$. If $5b9$ is divisible by 9, then $a + b$ equals:

(A) 2   (B) 4   (C) 6   (D) 8   (E) 9

2. An equivalent of the expression

$$\left(\frac{x^2 + 1}{x}\right)\left(\frac{y^2 + 1}{y}\right) + \left(\frac{x^2 - 1}{y}\right)\left(\frac{y^2 - 1}{x}\right), \quad xy \neq 0,$$

is:

(A) 1   (B) $2xy$ (C) $2x^2y^2 + 2$ (D) $2xy + 2/(xy)$
(E) $(2x/y) + (2y/x)$

3. The side of an equilateral triangle is s. A circle is inscribed in the triangle and a square is inscribed in the circle. The area of the square is:

(A) $\dfrac{s^2}{24}$ (B) $\dfrac{s^2}{6}$ (C) $\dfrac{s^2\sqrt{2}}{6}$ (D) $\dfrac{s^2\sqrt{3}}{6}$ (E) $\dfrac{s^2}{3}$

4. Given $(\log a)/p = (\log b)/q = (\log c)/r = \log x$, all logarithms to the same base and $x \neq 1$. If $b^2/(ac) = x^y$, then y is:

(A) $\dfrac{q^2}{p + r}$ (B) $\dfrac{p + r}{2q}$ (C) $2q - p - r$ (D) $2q - pr$

(E) $q^2 - pr$

5. A triangle is circumscribed about a circle of radius r inches. If the perimeter of the triangle is P inches and the area is K square inches, then P/K is:

(A) independent of the value of r (B) $\sqrt{2}/r$ (C) $2/\sqrt{r}$
(D) $2/r$ (E) $r/2$

6. If $f(x) = 4^x$ then $f(x + 1) - f(x)$ equals:

(A) 4   (B) $f(x)$ (C) $2f(x)$ (D) $3f(x)$ (E) $4f(x)$

7. If $a/b < -c/d$ where a, b, c, d are real numbers and $bd \neq 0$, then:

(A) a must be negative
(B) a must be positive
(C) a must not be zero
(D) a can be negative or zero, but not positive
(E) a can be positive, negative, or zero

8. To m ounces of an $m\%$ solution of acid, x ounces of water are added to yield an $(m - 10)\%$ solution. If $m > 25$, then x is:

(A) $\dfrac{10m}{m - 10}$ (B) $\dfrac{5m}{m - 10}$ (C) $\dfrac{m}{m - 10}$ (D) $\dfrac{5m}{m - 20}$

(E) not determined by the given information

9. Let K, in square units, be the area of a trapezoid such that the shorter base, the altitude, and the longer base, in that order, are in arithmetic progression. Then:

(A) K must be an integer   (B) K must be a rational fraction
(C) K must be an irrational number   (D) K must be an integer or a rational fraction   (E) taken alone neither (A) nor (B) nor (C) nor (D) is true

10. If $\dfrac{a}{10^x - 1} + \dfrac{b}{10^x + 2} = \dfrac{2 \cdot 10^x + 3}{(10^x - 1)(10^x + 2)}$ is an identity for positive rational values of x, then the value of $a - b$ is:

(A) $4/3$ (B) $5/3$ (C) 2 (D) $11/4$ (E) 3

11. If the perimeter of rectangle $ABCD$ is 20 inches, the least value of diagonal AC, in inches, is:

(A) 0 (B) $\sqrt{50}$ (C) 10 (D) $\sqrt{200}$ (E) none of these

12. If the (convex) area bounded by the x-axis and the lines $y = mx + 4$, $x = 1$, and $x = 4$ is 7, then m equals:

(A) $-1/2$ (B) $-2/3$ (C) $-3/2$ (D) -2 (E) none of these

13. A triangle ABC is to be constructed given side a (opposite angle A), angle B, and h_c, the altitude from C. If N is the number of noncongruent solutions, then N

(A) is 1   (B) is 2   (C) must be zero   (D) must be infinite
(E) must be zero or infinite

14. Let $f(t) = \dfrac{t}{1-t}$, $t \neq 1$. If $y = f(x)$, then x can be expressed as:

(A) $f\left(\dfrac{1}{y}\right)$ (B) $-f(y)$ (C) $-f(-y)$ (D) $f(-y)$ (E) $f(y)$

15. The difference in the areas of two similar triangles is 18 square feet, and the ratio of the larger area to the smaller is the square of an integer. The area of the smaller triangle, in square feet, is an integer, and one of its sides is 3 feet. The corresponding side of the larger triangle, in feet, is:

(A) 12   (B) 9   (C) $6\sqrt{2}$ (D) 6   (E) $3\sqrt{2}$

16. Let the product $(12)(15)(16)$, each factor written in base b, equal 3146 in base b. Let $s = 12 + 15 + 16$, each term expressed in base b. Then s, in base b, is:

(A) 43   (B) 44   (C) 45   (D) 46   (E) 47

17. If r_1 and r_2 are the distinct real roots of $x^2 + px + 8 = 0$, then it must follow that:

(A) $|r_1 + r_2| > 4\sqrt{2}$ (B) $|r_1| > 3$ or $|r_2| > 3$
(C) $|r_1| > 2$ and $|r_2| > 2$ (D) $r_1 < 0$ and $r_2 < 0$
(E) $|r_1 + r_2| < 4\sqrt{2}$

18. If $x^2 - 5x + 6 < 0$ and $P = x^2 + 5x + 6$ then

(A) P can take any real value   (B) $20 < P < 30$
(C) $0 < P < 20$ (D) $P < 0$
(E) $P > 30$

19. The area of a rectangle remains unchanged when it is made $2\frac{1}{2}$ inches longer and $\frac{2}{3}$ inch narrower, or when it is made $2\frac{1}{2}$ inches shorter and $\frac{4}{3}$ inch wider. Its area, in square inches, is:

(A) 30   (B) 80/3   (C) 24   (D) 45/2   (E) 20

20. A circle is inscribed in a square of side m, then a square is inscribed in that circle, then a circle is inscribed in the latter square, and so on. If S_n is the sum of the areas of the first n circles so inscribed, then, as n grows beyond all bounds, S_n approaches:

(A) $\dfrac{\pi m^2}{2}$ (B) $\dfrac{3\pi m^2}{8}$ (C) $\dfrac{\pi m^2}{3}$ (D) $\dfrac{\pi m^2}{4}$ (E) $\dfrac{\pi m^2}{8}$

Part 2

21. In right triangle ABC the hypotenuse $AB = 5$ and leg $AC = 3$. The bisector of angle A meets the opposite side in A_1. A second right triangle PQR is then constructed with hypotenuse $PQ = A_1B$ and leg $PR = A_1C$. If the bisector of angle P meets the opposite side in P_1, the length of PP_1 is:

(A) $\dfrac{3\sqrt{6}}{4}$ (B) $\dfrac{3\sqrt{5}}{4}$ (C) $\dfrac{3\sqrt{3}}{4}$ (D) $\dfrac{3\sqrt{2}}{2}$ (E) $\dfrac{15\sqrt{2}}{16}$

22. For natural numbers, when P is divided by D, the quotient is Q and the remainder is R. When Q is divided by D', the quotient is Q' and the remainder is R'. Then, when P is divided by DD', the remainder is:

(A) $R + R'D$ (B) $R' + RD$ (C) RR' (D) R (E) R'

23. If x is real and positive and grows beyond all bounds, then $\log_3 (6x - 5) - \log_3 (2x + 1)$ approaches:

(A) 0   (B) 1   (C) 3   (D) 4   (E) no finite number

24. The number of solution-pairs in positive integers of the equation $3x + 5y = 501$ is:

(A) 33   (B) 34   (C) 35   (D) 100   (E) none of these.

25. For every odd number $p > 1$ we have:

(A) $(p - 1)^{\frac{1}{2}(p-1)} - 1$ is divisible by $p - 2$
(B) $(p - 1)^{\frac{1}{2}(p-1)} + 1$ is divisible by p
(C) $(p - 1)^{\frac{1}{2}(p-1)}$ is divisible by p
(D) $(p - 1)^{\frac{1}{2}(p-1)} + 1$ is divisible by $p + 1$
(E) $(p - 1)^{\frac{1}{2}(p-1)} - 1$ is divisible by $p - 1$

26. If one uses only the tabular information $10^3 = 1000$, $10^4 = 10,000$, $2^{10} = 1024$, $2^{11} = 2048$, $2^{12} = 4096$, $2^{13} = 8192$, then the strongest statement one can make for $\log_{10} 2$ is that it lies between:

(A) $\frac{3}{10}$ and $\frac{4}{11}$ (B) $\frac{3}{10}$ and $\frac{4}{12}$ (C) $\frac{3}{10}$ and $\frac{4}{13}$
(D) $\frac{3}{10}$ and $\frac{40}{132}$ (E) $\frac{3}{11}$ and $\frac{40}{132}$

27. Two candles of the same length are made of different materials so that one burns out completely at a uniform rate in 3 hours and the

other in 4 hours. At what time P.M. should the candles be lighted so that, at 4 P.M., one stub is twice the length of the other?

(A) 1:24   (B) 1:28   (C) 1:36   (D) 1:40   (E) 1:48

28. Given the two hypotheses: I Some Mems are not Ens and II No Ens are Vees. If "some" means "at least one", we can conclude that:

(A) Some Mems are not Vees   (B) Some Vees are not Mems
(C) No Mem is a Vee   (D) Some Mems are Vees
(E) Neither (A) nor (B) nor (C) nor (D) is deducible from the given statements.

29. AB is a diameter of a circle. Tangents AD and BC are drawn so that AC and BD intersect in a point on the circle. If $AD = a$ and $BC = b$, $a \neq b$, the diameter of the circle is:

(A) $|a - b|$ (B) $\frac{1}{2}(a + b)$ (C) $\sqrt{ab}$

(D) $\dfrac{ab}{a + b}$ (E) $\dfrac{1}{2}\dfrac{ab}{a + b}$

30. A dealer bought n radios for d dollars, d a positive integer. He contributed two radios to a community bazaar at half their cost. The rest he sold at a profit of \$8 on each radio sold. If the overall profit was \$72, then the least possible value of n for the given information is:

(A) 18   (B) 16   (C) 15   (D) 12   (E) 11

Part 3

31. Let $D = a^2 + b^2 + c^2$, where a, b are consecutive integers and $c = ab$. Then $\sqrt{D}$ is:

(A) always an even integer
(B) sometimes an odd integer, sometimes not
(C) always an odd integer
(D) sometimes rational, sometimes not
(E) always irrational

32. In quadrilateral $ABCD$ with diagonals AC and BD intersecting at O, $BO = 4$, $OD = 6$, $AO = 8$, $OC = 3$, and $AB = 6$. The length of AD is:

(A) 9   (B) 10   (C) $6\sqrt{3}$ (D) $8\sqrt{2}$ (E) $\sqrt{166}$

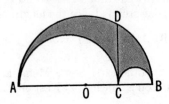

33. In this diagram semi-circles are constructed on diameters AB, AC and CB, so that they are mutually tangent. If $CD \perp AB$, then the ratio of the shaded area to the area of a circle with CD as *radius* is:

(A) 1:2   (B) 1:3   (C) $\sqrt{3}$:7   (D) 1:4   (E) $\sqrt{2}$:6

34. Points D, E, F are taken respectively on sides AB, BC, and CA of triangle ABC so that $AD:DB = BE:CE = CF:FA = 1:n$. The ratio of the area of triangle DEF to that of triangle ABC is:

(A) $\dfrac{n^2 - n + 1}{(n + 1)^2}$ (B) $\dfrac{1}{(n + 1)^2}$ (C) $\dfrac{2n^3}{(n + 1)^3}$

(D) $\dfrac{n^3}{(n + 1)^3}$ (E) $\dfrac{n(n - 1)}{n + 1}$

35. The roots of $64x^3 - 144x^2 + 92x - 15 = 0$ are in arithmetic progression. The difference between the largest and smallest roots is:

(A) 2   (B) 1   (C) 1/2   (D) 3/8   (E) 1/4

36. Given a geometric progression of five terms, each a positive integer less than 100. The sum of the five terms is 211. If S is the sum of those terms in the progression which are squares of integers, then S is:

(A) 0   (B) 91   (C) 133   (D) 195   (E) 211

37. Segments $AD = 10$, $BE = 6$, $CF = 24$ are drawn from the vertices of triangle ABC, each perpendicular to a straight line RS, not intersecting the triangle. Points D, E, F are the intersection points of RS with the perpendiculars. If x is the length of the perpendicular segment GH drawn to RS from the intersection point G of the medians of the triangle, then x is:

(A) 40/3   (B) 16   (C) 56/3   (D) 80/3   (E) undetermined

38. Given a set S consisting of two undefined elements "pib" and "maa", and the four postulates: P_1: Every pib is a collection of maas, P_2: Any two distinct pibs have one and only one maa in common, P_3: Every maa belongs to two and only two pibs, P_4: There are exactly four pibs.

Consider the three theorems: T_1: There are exactly six maas, T_2: There are exactly three maas in each pib, T_3: For each maa there is exactly one other maa not in the same pib with it. The theorems which are deducible from the postulates are:

(A) T_3 only   (B) T_2 and T_3 only
(C) T_1 and T_2 only   (D) T_1 and T_3 only
(E) all

39. Given the sets of consecutive integers $\{1\}$, $\{2, 3\}$, $\{4, 5, 6\}$, $\{7, 8, 9, 10\}$, $\cdots$, where each set contains one more element than the preceding one, and where the first element of each set is one more than the last element of the preceding set. Let S_n be the sum of the elements in the nth set. Then S_{21} equals:

(A) 1113   (B) 4641   (C) 5082   (D) 53361   (E) none of these

40. Located inside equilateral triangle ABC is a point P such that $PA = 6$, $PB = 8$, and $PC = 10$. To the nearest integer the area of triangle ABC is:

(A) 159   (B) 131   (C) 95   (D) 79   (E) 50

1968 Examination

Part 1

1. Let P units be the increase in the circumference of a circle resulting from an increase in π units in the diameter. Then P equals:

(A) $\dfrac{1}{\pi}$ (B) π (C) $\dfrac{\pi^2}{2}$ (D) π^2 (E) 2π

2. The real value of x such that 64^{x-1} divided by 4^{x-1} equals 256^{2x} is:

(A) $-\frac{2}{3}$ (B) $-\frac{1}{3}$ (C) 0 (D) $\frac{1}{4}$ (E) $\frac{3}{8}$

3. A straight line passing through the point $(0, 4)$ is perpendicular to the line $x - 3y - 7 = 0$. Its equation is:

(A) $y + 3x - 4 = 0$ (B) $y + 3x + 4 = 0$
(C) $y - 3x - 4 = 0$ (D) $3y + x - 12 = 0$
(E) $3y - x - 12 = 0$

4. Define an operation $*$ for positive real numbers as $a * b = \dfrac{ab}{a + b}$. Then $4 * (4 * 4)$ equals:

(A) $\frac{3}{4}$ (B) 1 (C) $\frac{4}{3}$ (D) 2 (E) $\frac{16}{3}$

5. If $f(n) = \frac{1}{3}n(n + 1)(n + 2)$, then $f(r) - f(r - 1)$ equals:

(A) $r(r + 1)$ (B) $(r + 1)(r + 2)$ (C) $\frac{1}{3}r(r + 1)$
(D) $\frac{1}{3}(r + 1)(r + 2)$ (E) $\frac{1}{3}r(r + 1)(2r + 1)$

6. Let side AD of convex quadrilateral $ABCD$ be extended through D, and let side BC be extended through C, to meet in point E. Let S represent the degree-sum of angles CDE and DCE, and let S' represent the degree-sum of angles BAD and ABC. If $r = S/S'$, then:

(A) $r = 1$ sometimes, $r > 1$ sometimes
(B) $r = 1$ sometimes, $r < 1$ sometimes
(C) $0 < r < 1$ (D) $r > 1$ (E) $r = 1$

7. Let O be the intersection point of medians AP and CQ of triangle ABC. If OQ is 3 inches, then OP, in inches, is:

   (A) 3   (B) $\frac{9}{2}$ (C) 6   (D) 9   (E) undetermined

8. A positive number is mistakenly divided by 6 instead of being multiplied by 6. Based on the correct answer, the error thus committed, to the nearest percent, is:

   (A) 100   (B) 97   (C) 83   (D) 17   (E) 3

9. The sum of the real values of x satisfying the equality $|x+2| = 2|x-2|$ is:

   (A) $\frac{1}{3}$ (B) $\frac{2}{3}$ (C) 6   (D) $6\frac{1}{3}$ (E) $6\frac{2}{3}$

10. Assume that, for a certain school, it is true that

    I: Some students are not honest.
    II: All fraternity members are honest.

    A necessary conclusion is:

    (A) Some students are fraternity members.
    (B) Some fraternity members are not students
    (C) Some students are not fraternity members
    (D) No fraternity member is a student
    (E) No student is a fraternity member.

Part 2

11. If an arc of 60° on circle I has the same length as an arc of 45° on circle II, the ratio of the area of circle I to that of circle II is:

    (A) 16:9   (B) 9:16   (C) 4:3   (D) 3:4
    (E) none of these

12. A circle passes through the vertices of a triangle with side-lengths $7\frac{1}{2}$, 10, $12\frac{1}{2}$. The radius of the circle is:

    (A) $\frac{15}{4}$ (B) 5   (C) $\frac{25}{4}$ (D) $\frac{35}{4}$ (E) $\dfrac{15\sqrt{2}}{2}$

13. If m and n are the roots of $x^2 + mx + n = 0$, $m \neq 0$, $n \neq 0$, then the sum of the roots is:

    (A) $-\frac{1}{2}$ (B) -1 (C) $\frac{1}{2}$ (D) 1   (E) undetermined

14. If x and y are non-zero numbers such that $x = 1 + \dfrac{1}{y}$ and $y = 1 + \dfrac{1}{x}$, then y equals

(A) $x - 1$ (B) $1 - x$ (C) $1 + x$ (D) $-x$ (E) x

15. Let P be the product of any three consecutive positive odd integers. The largest integer dividing all such P is:

(A) 15   (B) 6   (C) 5   (D) 3   (E) 1

16. If x is such that $\dfrac{1}{x} < 2$ and $\dfrac{1}{x} > -3$, then:

(A) $-\frac{1}{3} < x < \frac{1}{2}$ (B) $-\frac{1}{2} < x < 3$ (C) $x > \frac{1}{2}$
(D) $x > \frac{1}{2}$ or $-\frac{1}{3} < x < 0$ (E) $x > \frac{1}{2}$ or $x < -\frac{1}{3}$

17. Let $f(n) = \dfrac{x_1 + x_2 + \cdots + x_n}{n}$, where n is a positive integer. If $x_k = (-1)^k$, $k = 1, 2, \cdots, n$, the set of possible values of $f(n)$ is:

(A) $\{0\}$ (B) $\left\{\dfrac{1}{n}\right\}$ (C) $\left\{0, -\dfrac{1}{n}\right\}$ (D) $\left\{0, \dfrac{1}{n}\right\}$ (E) $\left\{1, \dfrac{1}{n}\right\}$

18. Side AB of triangle ABC has length 8 inches. Line DEF is drawn parallel to AB so that D is on segment AC, and E is on segment BC. Line AE extended bisects angle FEC. If DE has length 5 inches, then the length of CE, in inches, is:

(A) $\frac{51}{4}$ (B) 13   (C) $\frac{53}{4}$ (D) $\frac{40}{3}$ (E) $\frac{27}{2}$

19. Let n be the number of ways that 10 dollars can be changed into dimes and quarters, with at least one of each coin being used. Then n equals:

(A) 40   (B) 38   (C) 21   (D) 20   (E) 19

20. The measures of the interior angles of a convex polygon of n sides are in arithmetic progression. If the common difference is $5°$ and the largest angle is $160°$, then n equals:

(A) 9   (B) 10   (C) 12   (D) 16   (E) 32

Part 3

21. If $S = 1! + 2! + 3! + \cdots + 99!$, then the units' digit in the value of S is:†

    (A) 9   (B) 8   (C) 5   (D) 3   (E) 0

22. A segment of length 1 is divided into four segments. Then there exists a quadrilateral with the four segments as sides if and only if each segment is:

    (A) equal to $\frac{1}{4}$
    (B) equal to or greater than $\frac{1}{8}$ and less than $\frac{1}{2}$
    (C) greater than $\frac{1}{8}$ and less than $\frac{1}{2}$
    (D) greater than $\frac{1}{8}$ and less than $\frac{1}{4}$
    (E) less than $\frac{1}{2}$

23. If all the logarithms are real numbers, the equality

    $$\log (x + 3) + \log (x - 1) = \log (x^2 - 2x - 3)$$

    is satisfied for:

    (A) all real values of x
    (B) no real values of x
    (C) all real values of x except $x = 0$
    (D) no real values of x except $x = 0$
    (E) all real values of x except $x = 1$

24. A painting $18'' \times 24''$ is to be placed into a wooden frame with the longer dimension vertical. The wood at the top and bottom is twice as wide as the wood on the sides. If the frame area equals that of the painting itself, the ratio of the smaller to the larger dimension of the framed painting is:

    (A) 1:3   (B) 1:2   (C) 2:3   (D) 3:4   (E) 1:1

25. Ace runs with constant speed and Flash runs x times as fast, $x > 1$. Flash gives Ace a head start of y yards, and, at a given signal, they start off in the same direction. Then the number of yards Flash must run to catch Ace is:

    (A) xy (B) $\dfrac{y}{x + y}$ (C) $\dfrac{xy}{x - 1}$ (D) $\dfrac{x + y}{x + 1}$ (E) $\dfrac{x + y}{x - 1}$

† The symbol $n!$ denotes $1 \cdot 2 \cdot \cdots (n - 1)n$; thus $5! = 1 \cdot 2 \cdot 3 \cdot 4 \cdot 5 = 120$.

26. Let $S = 2 + 4 + 6 + \cdots + 2N$, where N is the smallest positive integer such that $S > 1,000,000$. Then the sum of the digits of N is:

(A) 27   (B) 12   (C) 6   (D) 2   (E) 1

27. Let $S_n = 1 - 2 + 3 - 4 + \cdots + (-1)^{n-1}n$, $n = 1, 2, \cdots$. Then $S_{17} + S_{33} + S_{50}$ equals:

(A) 0   (B) 1   (C) 2   (D) -1 (E) -2

28. If the arithmetic mean of a and b is double their geometric mean, with $a > b > 0$, then a possible value for the ratio a/b, to the nearest integer, is

(A) 5   (B) 8   (C) 11   (D) 14   (E) none of these

29. Given the three numbers x, $y = x^x$, $z = x^{(x^x)}$ with $.9 < x < 1.0$. Arranged in order of increasing magnitude, they are:

(A) x, z, y (B) x, y, z (C) y, x, z (D) y, z, x (E) z, x, y

30. Convex polygons P_1 and P_2 are drawn in the same plane with n_1 and n_2 sides, respectively, $n_1 \leq n_2$. If P_1 and P_2 do not have any line segment in common, then the maximum number of intersections of P_1 and P_2 is:

(A) $2n_1$ (B) $2n_2$ (C) n_1n_2 (D) $n_1 + n_2$ (E) none of these

Part 4

31. In this diagram, not drawn to scale, figures I and III are equilateral triangular regions with respective areas of $32\sqrt{3}$ and $8\sqrt{3}$ square inches. Figure II is a square region with area 32 sq. in. Let the length of segment AD be decreased by $12\frac{1}{2}\%$ of itself, while the lengths of AB and CD remain unchanged. The percent decrease in the area of the square is:

(A) $12\frac{1}{2}$ (B) 25   (C) 50   (D) 75   (E) $87\frac{1}{2}$

32. A and B move uniformly along two straight paths intersecting at right angles in point O. When A is at O, B is 500 yards short of O. In 2 minutes they are equidistant from O, and in 8 minutes more they are again equidistant from O. Then the ratio of A's speed to B's speed is:

(A) 4:5   (B) 5:6   (C) 2:3   (D) 5:8   (E) 1:2

33. A number N has three digits when expressed in base 7. When N is expressed in base 9 the digits are reversed. Then the middle digit is:

(A) 0   (B) 1   (C) 3   (D) 4   (E) 5

34. With 400 members voting the House of Representatives defeated a bill. A re-vote, with the same members voting, resulted in passage of the bill by twice the margin† by which it was originally defeated. The number voting for the bill on the re-vote was $\frac{12}{11}$ of the number voting against it originally. How many more members voted for the bill the second time than voted for it the first time?

(A) 75   (B) 60   (C) 50   (D) 45   (E) 20

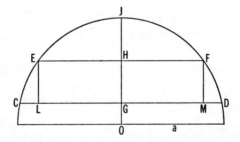

35. In this diagram the center of the circle is O, the radius is a inches, chord EF is parallel to chord CD, O, G, H, J are collinear, and G is the midpoint of CD. Let K (sq. in.) represent the area of trapezoid $CDFE$ and let R (sq. in.) represent the area of rectangle $ELMF$. Then, as CD and EF are translated upward so that OG increases toward the value a, while JH always equals HG, the ratio $K:R$ becomes arbitrarily close to:

(A) 0   (B) 1   (C) $\sqrt{2}$ (D) $\frac{1}{\sqrt{2}} + \frac{1}{2}$ (E) $\frac{1}{\sqrt{2}} + 1$

† In this context, margin of defeat (passage) is defined as the number of nays minus the number of ayes (nays − ayes).

1969 Examination

Part 1

1. When x is added to both the numerator and the denominator of the fraction a/b, $a \neq b$, $b \neq 0$, the value of the fraction is changed to c/d. Then x equals:

   (A) $\dfrac{1}{c-d}$ (B) $\dfrac{ad-bc}{c-d}$ (C) $\dfrac{ad-bc}{c+d}$ (D) $\dfrac{bc-ad}{c-d}$

   (E) $\dfrac{bc-ad}{c+d}$

2. If an item is sold for x dollars, there is a loss of 15% based on the cost.† If, however, the same item is sold for y dollars, there is a profit of 15% based on the cost.† The ratio $y:x$ is:

   (A) $23:17$ (B) $17y:23$ (C) $23x:17$
   (D) dependent upon the cost   (E) none of these.

3. If N, written in base 2, is 11000, the integer immediately preceding N, written in base 2, is:

   (A) 10001   (B) 10010   (C) 10011   (D) 10110   (E) 10111

4. Let a binary operation $*$ on ordered pairs of integers be defined by $(a, b) * (c, d) = (a - c, b + d)$. Then, if $(3, 2) * (0, 0)$ and $(x, y) * (3, 2)$ represent identical pairs, x equals:

   (A) -3 (B) 0 (C) 2 (D) 3 (E) 6

5. If a number N, $N \neq 0$, diminished by four times its reciprocal, equals a given real constant R, then, for this given R, the sum of all such possible values of N is:

   (A) $\dfrac{1}{R}$ (B) R (C) 4 (D) $\frac{1}{4}$ (E) $-R$

---

† $l\%$ loss based on cost means loss of $\dfrac{l}{100}\cdot$cost, $r\%$ profit based on cost means profit of $\dfrac{r}{100}\cdot$cost.

6. The area of the ring between two concentric circles is $12\frac{1}{2}\pi$ square inches. The length of a chord of the larger circle tangent to the smaller circle, in inches, is:

(A) $\dfrac{5}{\sqrt{2}}$ (B) 5   (C) $5\sqrt{2}$ (D) 10   (E) $10\sqrt{2}$

7. If the points $(1, y_1)$ and $(-1, y_2)$ lie on the graph of $y = ax^2 + bx + c$, and $y_1 - y_2 = -6$, then b equals:

(A) -3 (B) 0   (C) 3   (D) $\sqrt{ac}$ (E) $\dfrac{a+c}{2}$

8. Triangle ABC is inscribed in a circle. The measure of the non-overlapping minor arcs AB, BC, and CA are, respectively, $x + 75°$, $2x + 25°$, $3x - 22°$. Then one interior angle of the triangle, in degrees, is:

(A) $57\frac{1}{2}$ (B) 59   (C) 60   (D) 61   (E) 122

9. The arithmetic mean (ordinary average) of the fifty-two successive positive integers beginning with 2 is:

(A) 27   (B) $27\frac{1}{4}$ (C) $27\frac{1}{2}$ (D) 28   (E) $28\frac{1}{2}$

10. The number of points equidistant from a circle and two parallel tangents to the circle is:

(A) 0   (B) 2   (C) 3   (D) 4   (E) infinite

Part 2

11. Given points $P(-1, -2)$ and $Q(4, 2)$ in the xy-plane; point $R(1, m)$ is taken so that $PR + RQ$ is a minimum. Then m equals:

(A) $-\frac{3}{5}$ (B) $-\frac{2}{5}$ (C) $-\frac{1}{5}$ (D) $\frac{1}{5}$ (E) either $-\frac{1}{5}$ or $\frac{1}{5}$.

12. Let $F = \dfrac{6x^2 + 16x + 3m}{6}$ be the square of an expression which is linear in x. Then m has a particular value between:

(A) 3 and 4   (B) 4 and 5   (C) 5 and 6   (D) -4 and -3
(E) -6 and -5

13. A circle with radius r is contained within the region bounded by a circle with radius R. The area bounded by the larger circle is a/b

times the area of the region outside the smaller circle and inside the larger circle. Then $R:r$ equals:

(A) $\sqrt{a}:\sqrt{b}$ (B) $\sqrt{a}:\sqrt{a-b}$ (C) $\sqrt{b}:\sqrt{a-b}$
(D) $a:\sqrt{a-b}$ (E) $b:\sqrt{a-b}$

14. The complete set of x-values satisfying the inequality $\dfrac{x^2-4}{x^2-1}>0$

is the set of all x such that:

(A) $x>2$ or $x<-2$ or $-1<x<1$ (B) $x>2$ or $x<-2$
(C) $x>1$ or $x<-2$ (D) $x>1$ or $x<-1$
(E) x is any real number except 1 or -1

15. In a circle with center at O and radius r, chord AB is drawn with length equal to r(units). From O a perpendicular to AB meets AB at M. From M a perpendicular to OA meets OA at D. In terms of r the area of triangle MDA, in appropriate square units, is:

(A) $\dfrac{3r^2}{16}$ (B) $\dfrac{\pi r^2}{16}$ (C) $\dfrac{\pi r^2\sqrt{2}}{8}$ (D) $\dfrac{r^2\sqrt{3}}{32}$ (E) $\dfrac{r^2\sqrt{6}}{48}$

16. When $(a-b)^n$, $n\geq 2$, $ab\neq 0$, is expanded by the binomial theorem, it is found that, when $a=kb$, where k is a positive integer, the sum of the second and third terms is zero. Then n equals:

(A) $\tfrac{1}{2}k(k-1)$ (B) $\tfrac{1}{2}k(k+1)$ (C) $2k-1$ (D) $2k$
(E) $2k+1$

17. The equation $2^{2x}-8\cdot 2^x+12=0$ is satisfied by:

(A) $\log 3$ (B) $\tfrac{1}{2}\log 6$ (C) $1+\log \tfrac{3}{2}$ (D) $1+\dfrac{\log 3}{\log 2}$
(E) none of these

18. The number of points common to the graphs of

$$(x-y+2)(3x+y-4)=0 \text{ and } (x+y-2)(2x-5y+7)=0$$

is.

(A) 2   (B) 4   (C) 6   (D) 16   (E) infinite

19. The number of distinct ordered pairs (x, y), where x and y have positive integral values satisfying the equation $x^4y^4-10x^2y^2+9=0$, is:

(A) 0   (B) 3   (C) 4   (D) 12   (E) infinite

20. Let P equal the product of 3,659,893,456,789,325,678 and 342,973,489,379,256. The number of digits in P is:

(A) 36   (B) 35   (C) 34   (D) 33   (E) 32

Part 3

21. If the graph of $x^2 + y^2 = m$ is tangent to that of $x + y = \sqrt{2m}$, then:

(A) m must equal $\frac{1}{2}$ (B) m must equal $\frac{1}{\sqrt{2}}$

(C) m must equal $\sqrt{2}$ (D) m must equal 2
(E) m may be any non-negative real number

22. Let K be the measure of the area bounded by the x -axis, the line $x = 8$, and the curve defined by

$$f = \{(x,y) \mid y = x \text{ when } 0 \le x \le 5, \ y = 2x - 5 \text{ when } 5 \le x \le 8\}.$$

Then K is:

(A) 21.5   (B) 36.4   (C) 36.5   (D) 44
(E) less than 44 but arbitrarily close to it.

23. For any integer n greater than 1, the number of prime numbers greater than $n! + 1$ and less than $n! + n$ is:†

(A) 0   (B) 1   (C) $\frac{n}{2}$ for n even, $\frac{n+1}{2}$ for n odd
(D) $n - 1$ (E) n

24. When the natural numbers P and P' , with $P > P'$, are divided by the natural number D , the remainders are R and R' , respectively. When PP' and RR' are divided by D , the remainders are r and r' , respectively. Then:

(A) $r > r'$ always   (B) $r < r'$ always
(C) $r > r'$ sometimes, and $r < r'$ sometimes
(D) $r > r'$ sometimes, and $r = r'$ sometimes
(E) $r = r'$ always

25. If it is known that $\log_2 a + \log_2 b \ge 6$, then the least value that can be taken on by $a + b$ is:

(A) $2\sqrt{6}$ (B) 6   (C) $8\sqrt{2}$ (D) 16   (E) none of these.

† The symbol $n!$ denotes $1 \cdot 2 \cdot \cdots (n-1)n$; thus $5! = 1 \cdot 2 \cdot 3 \cdot 4 \cdot 5 = 120$.

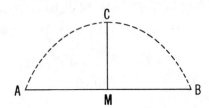

26. A parabolic arch has a height of 16 inches and a span of 40 inches. The height, in inches, of the arch at a point 5 inches from the center M is:

    (A) 1   (B) 15   (C) $15\frac{1}{3}$ (D) $15\frac{1}{2}$ (E) $15\frac{3}{4}$

27. A particle moves so that its speed for the second and subsequent miles varies inversely as the integral number of miles already traveled. For each subsequent mile the speed is constant. If the second mile is traversed in 2 hours, then the time, in hours, needed to traverse the nth mile is:

    (A) $\dfrac{2}{n-1}$ (B) $\dfrac{n-1}{2}$ (C) $\dfrac{2}{n}$ (D) $2n$ (E) $2(n-1)$

28. Let n be the number of points P interior to the region bounded by a circle with radius 1, such that the sum of the squares of the distances from P to the endpoints of a given diameter is 3. Then n is:

    (A) 0   (B) 1   (C) 2   (D) 4   (E) infinite

29. If $x = t^{1/(t-1)}$ and $y = t^{t/(t-1)}$, $t > 0$, $t \neq 1$, a relation between x and y is:

    (A) $y^x = x^{1/y}$ (B) $y^{1/x} = x^y$ (C) $y^x = x^y$ (D) $x^x = y^y$
    (E) none of these

30. Let P be a point of hypotenuse AB (or its extension) of isosceles right triangle ABC. Let $s = AP^2 + PB^2$. Then:

    (A) $s < 2CP^2$ for a finite number of positions of P
    (B) $s < 2CP^2$ for an infinite number of positions of P
    (C) $s = 2CP^2$ only if P is the midpoint of AB or an endpoint of AB
    (D) $s = 2CP^2$ always
    (E) $s > 2CP^2$ if P is a trisection point of AB

Part 4

31. Let $OABC$ be a unit square in the xy-plane with $O(0,0)$, $A(1,0)$, $B(1,1)$ and $C(0,1)$. Let $u = x^2 - y^2$ and $v = 2xy$ be a transformation of the xy-plane into the uv-plane. The transform (or image) of the square is:

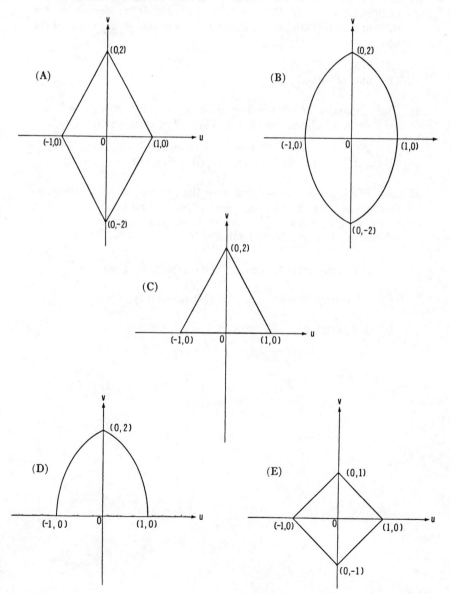

32. Let a sequence $\{u_n\}$ be defined by $u_1 = 5$ and the relation $u_{n+1} - u_n = 3 + 4(n - 1)$, $n = 1, 2, 3, \cdots$. If u_n is expressed as a polynomial in n, the algebraic sum of its coefficients is:

    (A) 3   (B) 4   (C) 5   (D) 6   (E) 11

33. Let S_n and T_n be the respective sums of the first n terms of two arithmetic series. If $S_n:T_n = (7n + 1):(4n + 27)$ for all n, the ratio of the eleventh term of the first series to the eleventh term of the second series is:

    (A) 4:3   (B) 3:2   (C) 7:4   (D) 78:71
    (E) undetermined

34. The remainder R obtained by dividing x^{100} by $x^2 - 3x + 2$ is a polynomial of degree less than 2. Then R may be written as:

    (A) $2^{100} - 1$ (B) $2^{100}(x - 1) - (x - 2)$ (C) $2^{100}(x - 3)$
    (D) $x(2^{100} - 1) + 2(2^{99} - 1)$ (E) $2^{100}(x + 1) - (x + 2)$

35. Let $L(m)$ be the x-coordinate of the left end point of the intersection of the graphs of $y = x^2 - 6$ and $y = m$, where $-6 < m < 6$. Let $r = [L(-m) - L(m)]/m$. Then, as m is made arbitrarily close to zero, the value of r is:

    (A) arbitrarily close to zero   (B) arbitrarily close to $\dfrac{1}{\sqrt{6}}$

    (C) arbitrarily close to $\dfrac{2}{\sqrt{6}}$ (D) arbitrarily large

    (E) undetermined

1970 Examination

Part 1

1. The fourth power of $\sqrt{1 + \sqrt{1 + \sqrt{1}}}$ is

   (A) $\sqrt{2} + \sqrt{3}$ (B) $\frac{1}{2}(7 + 3\sqrt{5})$ (C) $1 + 2\sqrt{3}$ (D) 3
   (E) $3 + 2\sqrt{2}$

2. A square and a circle have equal perimeters. The ratio of the area of the circle to the area of the square is

   (A) $4/\pi$ (B) $\pi/\sqrt{2}$ (C) $4/1$ (D) $\sqrt{2}/\pi$ (E) $\pi/4$

3. If $x = 1 + 2^p$ and $y = 1 + 2^{-p}$, then y in terms of x is

   (A) $\dfrac{x+1}{x-1}$ (B) $\dfrac{x+2}{x-1}$ (C) $\dfrac{x}{x-1}$ (D) $2 - x$

   (E) $\dfrac{x-1}{x}$

4. Let S be the set of all numbers which are the sum of the squares of three consecutive integers. Then we can say that

   (A) No member of S is divisible by 2
   (B) No member of S is divisible by 3 but some member is divisible by 11
   (C) No member of S is divisible by 3 or by 5
   (D) No member of S is divisible by 3 or by 7
   (E) None of these

5. If $f(x) = \dfrac{x^4 + x^2}{x + 1}$, then $f(i)$, where $i = \sqrt{-1}$, is equal to

   (A) $1 + i$ (B) 1 (C) -1 (D) 0 (E) $-1 - i$

6. The smallest value of $x^2 + 8x$ for real values of x is

   (A) -16.25 (B) -16 (C) -15 (D) -8
   (E) None of these

7. Inside square $ABCD$ with side s, quarter-circle arcs with radii s and centers at A and B are drawn. These arcs intersect at a point X inside the square. How far is X from side CD?

(A) $\frac{1}{2}s(\sqrt{3}+4)$ (B) $\frac{1}{2}s\sqrt{3}$ (C) $\frac{1}{2}s(1+\sqrt{3})$
(D) $\frac{1}{2}s(\sqrt{3}-1)$ (E) $\frac{1}{2}s(2-\sqrt{3})$

8. If $a = \log_8 225$ and $b = \log_2 15$, then

(A) $a = b/2$ (B) $a = 2b/3$ (C) $a = b$ (D) $b = a/2$
(E) $a = 3b/2$

9. Points P and Q are on line segment AB, and both points are on the same side of the midpoint of AB. Point P divides AB in the ratio 2:3, and Q divides AB in the ratio 3:4. If $PQ = 2$, then the length of segment AB is

(A) 12   (B) 28   (C) 70   (D) 75   (E) 105

10. Let $F = .48181\cdots$ be an infinite repeating decimal with the digits 8 and 1 repeating. When F is written as a fraction in lowest terms, the denominator exceeds the numerator by

(A) 13   (B) 14   (C) 29   (D) 57   (E) 126

Part 2

11. If two factors of $2x^3 - hx + k$ are $x + 2$ and $x - 1$, the value of $|2h - 3k|$ is

(A) 4   (B) 3   (C) 2   (D) 1   (E) 0

12. A circle with radius r is tangent to sides AB, AD, and CD of rectangle $ABCD$ and passes through the midpoint of diagonal AC. The area of the rectangle, in terms of r, is

(A) $4r^2$ (B) $6r^2$ (C) $8r^2$ (D) $12r^2$ (E) $20r^2$

13. Given the binary operation $*$ defined by $a * b = a^b$ for all positive numbers a and b. Then for all positive a, b, c, n, we have

(A) $a * b = b * a$ (B) $a * (b * c) = (a * b) * c$
(C) $(a * b^n) = (a * n) * b$ (D) $(a * b)^n = a * (bn)$
(E) None of these

14. Consider $x^2 + px + q = 0$, where p and q are positive numbers. If the roots of this equation differ by 1, then p equals

(A) $\sqrt{4q + 1}$ (B) $q - 1$ (C) $-\sqrt{4q + 1}$ (D) $q + 1$
(E) $\sqrt{4q - 1}$

15. Lines in the xy-plane are drawn through the point $(3, 4)$ and the trisection points of the line segment joining the points $(-4, 5)$ and $(5, -1)$. One of these lines has the equation

(A) $3x - 2y - 1 = 0$ (B) $4x - 5y + 8 = 0$
(C) $5x + 2y - 23 = 0$ (D) $x + 7y - 31 = 0$
(E) $x - 4y + 13 = 0$

16. If $F(n)$ is a function such that $F(1) = F(2) = F(3) = 1$, and such that $F(n + 1) = \dfrac{F(n) \cdot F(n - 1) + 1}{F(n - 2)}$ for $n \geq 3$, then $F(6)$ is equal to

(A) 2   (B) 3   (C) 7   (D) 11   (E) 26

17. If $r > 0$, then for all p and q such that $pq \neq 0$ and $pr > qr$, we have

(A) $-p > -q$ (B) $-p > q$ (C) $1 > -q/p$
(D) $1 < q/p$ (E) None of these

18. $\sqrt{3 + 2\sqrt{2}} - \sqrt{3 - 2\sqrt{2}}$ is equal to

(A) 2   (B) $2\sqrt{3}$ (C) $4\sqrt{2}$ (D) $\sqrt{6}$ (E) $2\sqrt{2}$

19. The sum of an infinite geometric series with common ratio r such that $|r| < 1$ is 15, and the sum of the squares of the terms of this series is 45. The first term of the series is

(A) 12   (B) 10   (C) 5   (D) 3   (E) 2

20. Lines HK and BC lie in a plane. M is the midpoint of line segment BC, and BH and CK are perpendicular to HK. Then we

(A) always have $MH = MK$ (B) always have $MH > BK$
(C) sometimes have $MH = MK$ but not always
(D) always have $MH > MB$ (E) always have $BH < BC$

Part 3

21. On an auto trip, the distance read from the instrument panel was 450 miles. With snow tires on for the return trip over the same route, the reading was 440 miles. Find, to the nearest hundredth of an inch, the increase in radius of the wheels if the original radius was 15 inches.

    (A) .33   (B) .34   (C) .35   (D) .38   (E) .66

22. If the sum of the first $3n$ positive integers is 150 more than the sum of the first n positive integers, then the sum of the first $4n$ positive integers is

    (A) 300   (B) 350   (C) 400   (D) 450   (E) 600

23. The number $10!$† (10 is written in base 10), when written in the base 12 system, ends with exactly k zeros. The value of k is

    (A) 1   (B) 2   (C) 3   (D) 4   (E) 5

24. An equilateral triangle and a regular hexagon have equal perimeters. If the area of the triangle is 2, then the area of the hexagon is

    (A) 2   (B) 3   (C) 4   (D) 6   (E) 12

25. For every real number x, let $[x]$ be the greatest integer which is less than or equal to x. If the postal rate for first class mail is six cents for every ounce or portion thereof, then the cost in cents of first-class postage on a letter weighing W ounces is always

    (A) $6W$ (B) $6[W]$ (C) $6([W]-1)$ (D) $6([W]+1)$
    (E) $-6[-W]$

26. The number of distinct points in the xy-plane common to the graphs of $(x+y-5)(2x-3y+5)=0$ and $(x-y+1)(3x+2y-12)=0$ is

    (A) 0   (B) 1   (C) 2   (D) 3   (E) 4   (F) infinite

27. In a triangle, the area is numerically equal to the perimeter. What is the radius of the inscribed circle?

    (A) 2   (B) 3   (C) 4   (D) 5   (E) 6

† The symbol $n!$ denotes $1 \cdot 2 \cdot \,\cdots\, \cdot (n-1)n$; thus $5! = 1 \cdot 2 \cdot 3 \cdot 4 \cdot 5 = 120$.

28. In triangle ABC, the median from vertex A is perpendicular to the median from vertex B. If the lengths of sides AC and BC are 6 and 7 respectively, then the length of side AB is

    (A) $\sqrt{17}$ (B) 4   (C) $4\frac{1}{2}$ (D) $2\sqrt{5}$ (E) $4\frac{1}{4}$

29. It is now between 10:00 and 11:00 o'clock, and six minutes from now, the minute hand of a watch will be exactly opposite the place where the hour hand was three minutes ago. What is the exact time now?

    (A) $10:05\frac{5}{11}$ (B) $10:07\frac{1}{2}$ (C) 10:10   (D) 10:15
    (E) $10:17\frac{1}{2}$

30. In the accompanying figure, segments AB and CD are parallel, the measure of angle D is twice that of angle B, and the measures of segments AD and CD are a and b respectively. Then the measure of AB is equal to

    (A) $\frac{1}{2}a + 2b$ (B) $\frac{3}{2}b + \frac{3}{4}a$ (C) $2a - b$ (D) $4b - \frac{1}{2}a$
    (E) $a + b$

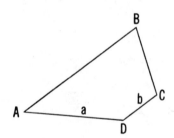

Part 4

31. If a number is selected at random from the set of all five-digit numbers in which the sum of the digits is equal to 43, what is the probability that this number will be divisible by 11?

    (A) 2/5   (B) 1/5   (C) 1/6   (D) 1/11   (E) 1/15

32. A and B travel around a circular track at uniform speeds in opposite directions, starting from diametrically opposite points. If they start at the same time, meet first after B has travelled 100 yards, and meet a second time 60 yards before A completes one lap, then the circumference of the track in yards is

    (A) 400   (B) 440   (C) 480   (D) 560   (E) 880

33. Find the sum of the digits of all the numbers in the sequence 1, 2, 3, 4, ···, 10000.

(A) 180,001   (B) 154,756   (C) 45,001   (D) 154,755
(E) 270,001

34. The greatest integer that will divide 13,511, 13,903 and 14,589 and leave the same remainder is

(A) 28   (B) 49   (C) 98
(D) an odd multiple of 7 greater than 49
(E) an even multiple of 7 greater than 98

35. A retiring employee receives an annual pension proportional to the square root of the number of years of his service. Had he served a years more, his pension would have been p dollars greater, whereas, had he served b years more $(b \neq a)$, his pension would have been q dollars greater than the original annual pension. Find his annual pension in terms of a, b, p, and q.

(A) $\dfrac{p^2 - q^2}{2(a - b)}$ (B) $\dfrac{(p - q)^2}{2\sqrt{ab}}$ (C) $\dfrac{ap^2 - bq^2}{2(ap - bq)}$ (D) $\dfrac{aq^2 - bp^2}{2(bp - aq)}$

(E) $\sqrt{(a - b)(p - q)}$

1971 Examination

Part 1

1. The number of digits in the number $N = 2^{12} \times 5^8$ is

   (A) 9   (B) 10   (C) 11   (D) 12   (E) 20

2. If b men take c days to lay f bricks, then the number of days it will take c men working at the same rate to lay b bricks, is

   (A) fb^2 (B) b/f^2 (C) f^2/b (D) b^2/f (E) f/b^2

3. If the point $(x, -4)$ lies on the straight line joining the points $(0, 8)$ and $(-4, 0)$ in the xy-plane, then x is equal to

   (A) -2 (B) 2   (C) -8 (D) 6   (E) -6

4. After simple interest for two months at 5% per annum was credited, a Boy Scout Troop had a total of $255.31 in the Council Treasury. The interest credited was a number of dollars plus the following number of cents

   (A) 11   (B) 12   (C) 13   (D) 21   (E) 31

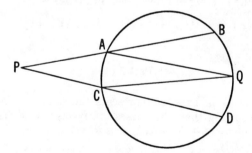

5. Points A, B, Q, D, and C lie on the circle shown and the measures of arcs $\overset{\frown}{BQ}$ and $\overset{\frown}{QD}$ are 42° and 38° respectively. The sum of the measures of angles P and Q is

   (A) 80°   (B) 62°   (C) 40°   (D) 46°   (E) None of these

6. Let $*$ be a symbol denoting the binary operation on the set S of all non-zero real numbers as follows: For any two numbers a and

b of S, $a * b = 2ab$. Then the one of the following statements which is not true, is

(A) $*$ is commutative over S (B) $*$ is associative over S
(C) $\frac{1}{2}$ is an identity element for $*$ in S
(D) Every element of S has an inverse for $*$
(E) $1/2a$ is an inverse for $*$ of the element a of S

7. $2^{-(2k+1)} - 2^{-(2k-1)} + 2^{-2k}$ is equal to

(A) 2^{-2k} (B) $2^{-(2k-1)}$ (C) $-2^{-(2k+1)}$ (D) 0 (E) 2

8. The solution set of $6x^2 + 5x < 4$ is the set of all values of x such that

(A) $-2 < x < 1$ (B) $-\frac{4}{3} < x < \frac{1}{2}$ (C) $-\frac{1}{2} < x < \frac{4}{3}$
(D) $x < \frac{1}{2}$ or $x > -\frac{4}{3}$ (E) $x < -\frac{4}{3}$ or $x > \frac{1}{2}$

9. An uncrossed belt is fitted without slack around two circular pulleys with radii of 14 inches and 4 inches. If the distance between the points of contact of the belt with the pulleys is 24 inches, then the distance between the centers of the pulleys in inches is

(A) 24 (B) $2\sqrt{119}$ (C) 25 (D) 26 (E) $4\sqrt{35}$

10. Each of a group of 50 girls is blonde or brunette and is blue or brown-eyed. If 14 are blue-eyed blondes, 31 are brunettes, and 18 are brown-eyed, then the number of brown-eyed brunettes is

(A) 5 (B) 7 (C) 9 (D) 11 (E) 13

Part 2

11. The numeral 47 in base a represents the same number as 74 in base b. Assuming that both bases are positive integers, the least possible value for $a + b$ written as a Roman numeral, is

(A) XIII   (B) XV   (C) XXI   (D) XXIV   (E) XVI

12. For each integer $N > 1$, there is a mathematical system in which two or more integers are defined to be congruent if they leave the same non-negative remainder when divided by N. If 69, 90, and 125 are congruent in one such system, then in that same system, 81 is congruent to

(A) 3 (B) 4 (C) 5 (D) 7 (E) 8

13. If $(1.0025)^{10}$ is evaluated correct to 5 decimal places, then the digit in the fifth decimal place is

(A) 0   (B) 1   (C) 2   (D) 5   (E) 8

14. The number $(2^{48} - 1)$ is exactly divisible by two numbers between 60 and 70. These numbers are

(A) 61, 63   (B) 61, 65   (C) 63, 65   (D) 63, 67   (E) 67, 69

15. An aquarium on a level table has rectangular faces and is 10 inches wide and 8 inches high. When it was tilted, the water in it just covered an $8'' \times 10''$ end but only three-fourths of the rectangular bottom. The depth of the water when the bottom was again made level was

(A) $2\frac{1}{2}''$ (B) $3''$ (C) $3\frac{1}{4}''$ (D) $3\frac{1}{2}''$ (E) $4''$

16. After finding the average of 35 scores, a student carelessly included the average with the 35 scores and found the average of these 36 numbers. The ratio of the second average to the true average was

(A) 1:1   (B) 35:36   (C) 36:35   (D) 2:1   (E) None of these

17. A circular disk is divided by $2n$ equally spaced radii $(n > 0)$ and one secant line. The maximum number of non-overlapping areas into which the disk can be divided is

(A) $2n + 1$ (B) $2n + 2$ (C) $3n - 1$ (D) $3n$ (E) $3n + 1$

18. The current in a river is flowing steadily at 3 miles per hour. A motor boat which travels at a constant rate in still water goes downstream 4 miles and then returns to its starting point. The trip takes one hour, excluding the time spent in turning the boat around. The ratio of the downstream to the upstream rate is

(A) 4:3   (B) 3:2   (C) 5:3   (D) 2:1   (E) 5:2

19. If the line $y = mx + 1$ intersects the ellipse $x^2 + 4y^2 = 1$ exactly once, then the value of m^2 is

(A) $\frac{1}{2}$ (B) $\frac{2}{3}$ (C) $\frac{3}{4}$ (D) $\frac{4}{5}$ (E) $\frac{5}{6}$

20. The sum of the squares of the roots of the equation $x^2 + 2hx = 3$ is 10. The absolute value of h is equal to

(A) -1 (B) $\frac{1}{2}$ (C) $\frac{3}{2}$ (D) 2   (E) None of these

Part 3

21. If $\log_2 (\log_3 (\log_4 x)) = \log_3 (\log_4 (\log_2 y)) = \log_4 (\log_2 (\log_3 z)) = 0$, then the sum $x + y + z$ is equal to

(A) 50   (B) 58   (C) 89   (D) 111   (E) 1296

22. If w is one of the imaginary roots of the equation $x^3 = 1$, then the product $(1 - w + w^2)(1 + w - w^2)$ is equal to

(A) 4   (B) w (C) 2   (D) w^2 (E) 1

23. Teams A and B are playing a series of games. If the odds for either team to win any game are even and Team A must win two or Team B three games to win the series, then the odds favoring Team A to win the series are

(A) 11 to 5   (B) 5 to 2   (C) 8 to 3   (D) 3 to 2   (E) 13 to 6

$$1$$
$$1\ 1$$
$$1\ 2\ 1$$
$$1\ 3\ 3\ 1$$
$$1\ 4\ 6\ 4\ 1$$
$$\text{etc.}$$

24. Pascal's triangle is an array of positive integers (see figure), in which the first row is 1, the second row is two 1's, each row begins and ends with 1, and the kth number in any row when it is not 1, is the sum of the kth and $(k - 1)$th numbers in the immediately preceding row. The quotient of the number of numbers in the first n rows which are not 1's and the number of 1's is

(A) $\dfrac{n^2 - n}{2n - 1}$ (B) $\dfrac{n^2 - n}{4n - 2}$ (C) $\dfrac{n^2 - 2n}{2n - 1}$ (D) $\dfrac{n^2 - 3n + 2}{4n - 2}$

(E) None of these

25. A teen age boy wrote his own age after his father's. From this new four place number he subtracted the absolute value of the difference of their ages to get 4,289. The sum of their ages was

(A) 48   (B) 52   (C) 56   (D) 59   (E) 64

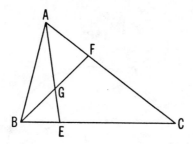

26. In triangle ABC, point F divides side AC in the ratio $1:2$. Let E be the point of intersection of side BC and AG where G is the midpoint of BF. Then point E divides side BC in the ratio

    (A) 1:4   (B) 1:3   (C) 2:5   (D) 4:11   (E) 3:8

27. A box contains chips, each of which is red, white, or blue. The number of blue chips is at least half the number of white chips, and at most one third the number of red chips. The number which are white or blue is at least 55. The minimum number of red chips is

    (A) 24   (B) 33   (C) 45   (D) 54   (E) 57

28. Nine lines parallel to the base of a triangle divide the other sides each into 10 equal segments and the area into 10 distinct parts. If the area of the largest of these parts is 38, then the area of the original triangle is

    (A) 180   (B) 190   (C) 200   (D) 210   (E) 240

29. Given the progression $10^{1/11}$, $10^{2/11}$, $10^{3/11}$, $10^{4/11}$, $\cdots$, $10^{n/11}$. The least positive integer n such that the product of the first n terms of the progression exceeds $100,000$ is

    (A) 7   (B) 8   (C) 9   (D) 10   (E) 11

30. Given the linear fractional transformation of x into $f_1(x) = \dfrac{2x-1}{x+1}$.

    Define

    $$f_{n+1}(x) = f_1(f_n(x)) \quad \text{for} \quad n = 1, 2, 3, \cdots.$$

    It can be shown that $f_{35} = f_5$; it follows that $f_{28}(x)$ is

    (A) x (B) $\dfrac{1}{x}$ (C) $\dfrac{x-1}{x}$ (D) $\dfrac{1}{1-x}$ (E) None of these

Part 4

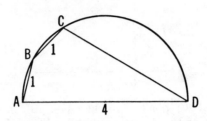

31. Quadrilateral $ABCD$ is inscribed in a circle with side AD, a diameter of length 4. If sides AB and BC each have length 1, then side CD has length

    (A) $\dfrac{7}{2}$ (B) $\dfrac{5\sqrt{2}}{2}$ (C) $\sqrt{11}$ (D) $\sqrt{13}$ (E) $2\sqrt{3}$

32. If $s = (1 + 2^{-1/32})(1 + 2^{-1/16})(1 + 2^{-1/8})(1 + 2^{-1/4})(1 + 2^{-1/2})$, then s is equal to

    (A) $\frac{1}{2}(1 - 2^{-1/32})^{-1}$ (B) $(1 - 2^{-1/32})^{-1}$ (C) $1 - 2^{-1/32}$
    (D) $\frac{1}{2}(1 - 2^{-1/32})$ (E) $\frac{1}{2}$

33. If P is the product of n quantities in geometric progression, S their sum, and S' the sum of their reciprocals, then P in terms of S, S', and n is

    (A) $(SS')^{n/2}$ (B) $(S/S')^{n/2}$ (C) $(SS')^{n-2}$ (D) $(S/S')^{n}$
    (E) $(S'/S)^{(n-1)/2}$

34. An ordinary clock in a factory is running slow so that the minute hand passes the hour hand at the usual dial positions (12 o'clock, etc.) but only every 69 minutes. At time and one-half for overtime, the extra pay to which a \$4.00 per hour worker should be entitled after working a normal 8 hour day by that slow running clock, is

    (A) \$2.30   (B) \$2.60   (C) \$2.80   (D) \$3.00   (E) \$3.30

35. Each circle in an infinite sequence with decreasing radii is tangent externally to the one following it and to both sides of a given right angle. The ratio of the area of the first circle to the sum of areas of all other circles in the sequence is

    (A) $(4 + 3\sqrt{2}):4$ (B) $9\sqrt{2}:2$ (C) $(16 + 12\sqrt{2}):1$
    (D) $(2 + 2\sqrt{2}):1$ (E) $(3 + 2\sqrt{2}):1$

1972 Examination

Part 1

1. The lengths in inches of the three sides of each of four triangles I, II, III, and IV are as follows:

   I 3, 4, and 5   III 7, 24, and 25
   II 4, $7\frac{1}{2}$, and $8\frac{1}{2}$ IV $3\frac{1}{2}$, $4\frac{1}{2}$, and $5\frac{1}{2}$.

   Of these four given triangles, the only right triangles are

   (A) I and II   (B) I and III   (C) I and IV   (D) I, II, and III
   (E) I, II, and IV

2. If a dealer could get his goods for 8% less while keeping his selling price fixed, his profit, based on cost,† would be increased to $(x + 10)\%$ from his present profit of $x\%$ which is

   (A) 12%   (B) 15%   (C) 30%   (D) 50%   (E) 75%

3. If $x = \dfrac{1 - i\sqrt{3}}{2}$, where $i = \sqrt{-1}$, then $\dfrac{1}{x^2 - x}$ is equal to

   (A) -2 (B) -1 (C) $1 + i\sqrt{3}$ (D) 1   (E) 2

4. The number of solutions to $\{1, 2\} \subseteq X \subseteq \{1, 2, 3, 4, 5\}$, where X is a set, is

   (A) 2   (B) 4   (C) 6   (D) 8   (E) None of these

5. From among $2^{1/2}$, $3^{1/3}$, $8^{1/8}$, $9^{1/9}$ those which have the greatest and the next to the greatest values, in that order, are

   (A) $3^{1/3}$, $2^{1/2}$ (B) $3^{1/3}$, $8^{1/8}$ (C) $3^{1/3}$, $9^{1/9}$ (D) $8^{1/8}$, $9^{1/9}$
   (E) None of these

6. If $3^{2x} + 9 = 10(3^x)$, then the value of $x^2 + 1$ is

   (A) 1 only   (B) 5 only   (C) 1 or 5   (D) 2   (E) 10

7. If $yz:zx:xy = 1:2:3$, then $\dfrac{x}{yz} : \dfrac{y}{zx}$ is equal to

   (A) 3:2   (B) 1:2   (C) 1:4   (D) 2:1   (E) 4:1

† $r\%$ profit based on cost means $\dfrac{r}{100}\cdot$cost.

8. If $|x - \log y| = x + \log y$, where x and $\log y$ are real, then

(A) $x = 0$ (B) $y = 1$ (C) $x = 0$ and $y = 1$
(D) $x(y - 1) = 0$ (E) None of these

9. Ann and Sue bought identical boxes of stationery. Ann used hers to write 1-sheet letters and Sue used hers to write 3-sheet letters. Ann used all the envelopes and had 50 sheets of paper left, while Sue used all of the sheets of paper and had 50 envelopes left. The number of sheets of paper in each box was

(A) 150   (B) 125   (C) 120   (D) 100   (E) 80

10. For x real, the inequality $1 \le |x - 2| \le 7$ is equivalent to

(A) $x \le 1$ or $x \ge 3$ (B) $1 \le x \le 3$ (C) $-5 \le x \le 9$
(D) $-5 \le x \le 1$ or $3 \le x \le 9$
(E) $-6 \le x \le 1$ or $3 \le x \le 10$

Part 2

11. The value(s) of y for which the pair of equations

$$x^2 + y^2 - 16 = 0 \text{ and } x^2 - 3y + 12 = 0$$

may have a real common solution, are

(A) 4 only   (B) $-7, 4$ (C) $0, 4$ (D) no y (E) all y

12. The number of cubic feet in the volume of a cube is the same as the number of square inches in its surface area. The length of the edge expressed in feet is

(A) 6   (B) 864   (C) 1728   (D) 6×1728 (E) 2304

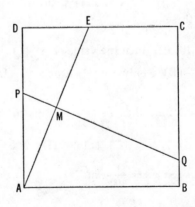

13. Inside square $ABCD$ (see figure) with sides of length 12 inches, segment AE is drawn, where E is the point on DC which is 5 inches from D. The perpendicular bisector of AE is drawn and intersects AE, AD, and BC at points M, P, and Q respectively. The ratio of segment PM to MQ is

   (A) 5:12   (B) 5:13   (C) 5:19   (D) 1:4   (E) 5:21

14. A triangle has angles of 30° and 45°. If the side opposite the 45° angle has length 8, then the side opposite the 30° angle has length

   (A) 4   (B) $4\sqrt{2}$ (C) $4\sqrt{3}$ (D) $4\sqrt{6}$ (E) 6

15. A contractor estimated that one of his two bricklayers would take 9 hours to build a certain wall and the other 10 hours. However, he knew from experience that when they worked together, their combined output fell by 10 bricks per hour. Being in a hurry, he put both men on the job and found that it took exactly 5 hours to build the wall. The number of bricks in the wall was

   (A) 500   (B) 550   (C) 900   (D) 950   (E) 960

16. There are two positive numbers that may be inserted between 3 and 9 such that the first three are in geometric progression while the last three are in arithmetic progression. The sum of those two positive numbers is

   (A) $13\frac{1}{2}$ (B) $11\frac{1}{4}$ (C) $10\frac{1}{2}$ (D) 10   (E) $9\frac{1}{2}$

17. A piece of string is cut in two at a point selected at random. The probability that the longer piece is at least x times as large as the shorter piece (where $x \geq 1$) is

   (A) $\dfrac{1}{2}$ (B) $\dfrac{2}{x}$ (C) $\dfrac{1}{x+1}$ (D) $\dfrac{1}{x}$ (E) $\dfrac{2}{x+1}$

18. Let $ABCD$ be a trapezoid with the measure of base AB twice that of base DC, and let E be the point of intersection of the diagonals. If the measure of diagonal AC is 11, then that of segment EC is equal to

   (A) $3\frac{2}{3}$ (B) $3\frac{3}{4}$ (C) 4   (D) $3\frac{1}{2}$ (E) 3

19. The sum of the first n terms of the sequence

   $$1, (1 + 2), (1 + 2 + 2^2), \cdots (1 + 2 + 2^2 + \cdots + 2^{n-1})$$

   in terms of n is

   (A) 2^n (B) $2^n - n$ (C) $2^{n+1} - n$ (D) $2^{n+1} - n - 2$ (E) $n \cdot 2^n$

20. If $\tan x = \dfrac{2ab}{a^2 - b^2}$, where $a > b > 0$ and $0° < x < 90°$, then

sin x is equal to

(A) $\dfrac{a}{b}$ (B) $\dfrac{b}{a}$ (C) $\dfrac{\sqrt{a^2 - b^2}}{2a}$ (D) $\dfrac{\sqrt{a^2 - b^2}}{2ab}$

(E) $\dfrac{2ab}{a^2 + b^2}$

Part 3

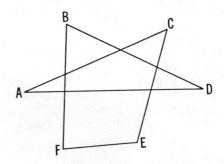

21. If the sum of the measures in degrees of angles A, B, C, D, E, and F in the figure is $90n$, then n is equal to

(A) 2   (B) 3   (C) 4   (D) 5   (E) 6

22. If $a \pm bi$ ($b \neq 0$, $i = \sqrt{-1}$) are imaginary roots of the equation $x^3 + qx + r = 0$, where a, b, q, and r are real numbers, then q in terms of a and b is

(A) $a^2 + b^2$ (B) $2a^2 - b^2$ (C) $b^2 - a^2$ (D) $b^2 - 2a^2$
(E) $b^2 - 3a^2$

23. The radius of the smallest circle containing the symmetric figure composed of the 3 unit squares is

(A) $\sqrt{2}$ (B) $\sqrt{1.25}$ (C) 1.25 (D) $\dfrac{5\sqrt{17}}{16}$ (E) None of these

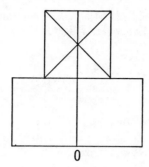

0

24. A man walked a certain distance at a constant rate. If he had gone $\frac{1}{2}$ mile per hour faster, he would have walked the distance in four-fifths of the time; if he had gone $\frac{1}{2}$ mile per hour slower, he would have been $2\frac{1}{2}$ hours longer on the road. The distance in miles he walked was

(A) $13\frac{1}{2}$ (B) 15   (C) $17\frac{1}{2}$ (D) 20   (E) 25

25. Inscribed in a circle is a quadrilateral having sides of lengths 25, 39, 52, and 60 taken consecutively. The diameter of this circle has length

(A) 62   (B) 63   (C) 65   (D) 66   (E) 69

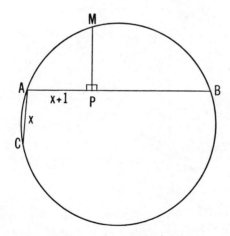

26. In the circle above, M is the mid-point of arc CAB, and segment MP is perpendicular to chord AB at P. If the measure of chord AC is x and that of segment AP is $(x + 1)$, then segment PB has measure equal to

(A) $3x + 2$ (B) $3x + 1$ (C) $2x + 3$ (D) $2x + 2$ (E) $2x + 1$

27. If the area of $\triangle ABC$ is 64 square inches and the geometric mean (mean proportional) between sides AB and AC is 12 inches, then $\sin A$ is equal to

   (A) $\dfrac{\sqrt{3}}{2}$ (B) $\dfrac{3}{5}$ (C) $\dfrac{4}{5}$ (D) $\dfrac{8}{9}$ (E) $\dfrac{15}{17}$

28. A circular disc with diameter D is placed on an 8×8 checkerboard with width D so that the centers coincide. The number of checkerboard squares which are completely covered by the disc is

   (A) 48   (B) 44   (C) 40   (D) 36   (E) 32

29. If $f(x) = \log\left(\dfrac{1+x}{1-x}\right)$ for $-1 < x < 1$, then $f\left(\dfrac{3x + x^3}{1 + 3x^2}\right)$ in terms of $f(x)$ is

   (A) $-f(x)$ (B) $2f(x)$ (C) $3f(x)$ (D) $[f(x)]^2$
   (E) $[f(x)]^3 - f(x)$

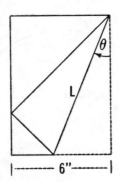

30. A rectangular piece of paper 6 inches wide is folded as in the diagram so that one corner touches the opposite side. The length in inches of the crease L in terms of angle θ is

   (A) $3\sec^2\theta\csc\theta$ (B) $6\sin\theta\sec\theta$ (C) $3\sec\theta\csc\theta$
   (D) $6\sec\theta\csc^2\theta$ (E) None of these

Part 4

31. When the number 2^{1000} is divided by 13, the remainder in the division is

    (A) 1   (B) 2   (C) 3   (D) 7   (E) 11

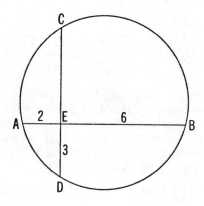

32. Chords AB and CD in the circle (see figure) intersect at E and are perpendicular to each other. If segments AE, EB, and ED have measures 2, 6, and 3 respectively, then the length of the diameter of the circle is

    (A) $4\sqrt{5}$ (B) $\sqrt{65}$ (C) $2\sqrt{17}$ (D) $3\sqrt{7}$ (E) $6\sqrt{2}$

33. The minimum value of the quotient of a (base ten) number of three different nonzero digits divided by the sum of its digits is

    (A) 9.7   (B) 10.1   (C) 10.5   (D) 10.9   (E) 20.5

34. Three times Dick's age plus Tom's age equals twice Harry's age. Double the cube of Harry's age is equal to three times the cube of Dick's age added to the cube of Tom's age. Their respective ages are relatively prime to each other. The sum of the squares of their ages is

    (A) 42   (B) 46   (C) 122   (D) 290   (E) 326

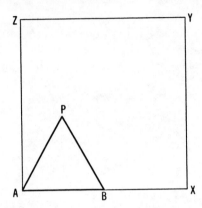

35. Equilateral triangle ABP (see figure) with side AB of length 2 inches is placed inside square $AXYZ$ with side of length 4 inches so that B is on side AX. The triangle is rotated clockwise about B, then P, and so on along the sides of the square until P, A, and B all return to their original positions.† The length of the path in inches traversed by vertex P is equal to

   (A) $20\pi/3$ (B) $32\pi/3$ (C) 12π (D) $40\pi/3$ (E) 15π

---

† The original wording of this problem was somewhat different. The reasons for the change are explained in the Comment, see p. 180.

II

Answer Keys

1966 Answers

1. C	9. A	17. C	25. D	33. D
2. E	10. E	18. A	26. C	34. B
3. D	11. C	19. B	27. A	35. C
4. B	12. E	20. C	28. B	36. E
5. A	13. E	21. E	29. B	37. C
6. C	14. C	22. A	30. D	38. D
7. A	15. D	23. A	31. D	39. E
8. B	16. B	24. B	32. B	40. A

1967 Answers

1. C	9. E	17. A	25. A	33. D
2. D	10. A	18. B	26. C	34. A
3. B	11. B	19. E	27. C	35. B
4. C	12. B	20. A	28. E	36. C
5. D	13. E	21. B	29. C	37. A
6. D	14. C	22. A	30. D	38. E
7. E	15. D	23. B	31. C	39. B
8. A	16. B	24. A	32. E	40. D

1968 Answers

1. D	8. B	15. D	22. E	29. A
2. B	9. E	16. E	23. B	30. A
3. A	10. C	17. C	24. C	31. D
4. C	11. B	18. D	25. C	32. C
5. A	12. C	19. E	26. E	33. A
6. E	13. B	20. A	27. B	34. B
7. E	14. E	21. D	28. D	35. D

1969 Answers

1. B	8. D	15. D	22. C	29. C
2. A	9. C	16. E	23. A	30. D
3. E	10. C	17. D	24. E	31. D
4. E	11. B	18. B	25. D	32. C
5. B	12. A	19. B	26. B	33. A
6. C	13. B	20. C	27. E	34. B
7. A	14. A	21. E	28. E	35. B

1970 Answers

1. E	8. B	15. E	22. A	29. D
2. A	9. C	16. C	23. D	30. E
3. C	10. D	17. E	24. B	31. B
4. B	11. E	18. A	25. E	32. C
5. D	12. C	19. C	26. B	33. A
6. B	13. D	20. A	27. A	34. C
7. E	14. A	21. B	28. A	35. D

1971 Answers

1. B	8. B	15. B	22. A	29. E
2. D	9. D	16. A	23. A	30. D
3. E	10. E	17. E	24. D	31. A
4. A	11. D	18. D	25. D	32. A
5. C	12. B	19. C	26. B	33. B
6. E	13. E	20. E	27. E	34. B
7. C	14. C	21. C	28. C	35. C

1972 Answers

1. D	6. C	11. A	16. B	21. C	26. E	31. C
2. B	7. E	12. B	17. E	22. E	27. D	32. B
3. B	8. D	13. C	18. A	23. D	28. E	33. C
4. D	9. A	14. B	19. D	24. B	29. C	34. A
5. A	10. D	15. C	20. E	25. C	30. A	35. D

III

Solutions†

1966 Solutions

Part 1

1. (C) We have $(3x - 4) = k(y + 15)$, where the constant ratio $k = \frac{1}{9}$ is determined by replacing (x, y) by $(2, 3)$. The relation $(3x - 4) = \frac{1}{9}(y + 15)$ now yields $x = \frac{7}{3}$ when we set $y = 12$.

2. (E) If b and h denote the base and altitude of the triangle, then after the ten percent changes, the area becomes

$$\tfrac{1}{2}(1.1b)(.9h) = .99(\tfrac{1}{2}bh)$$

which is a 1% decrease from the original area $\frac{1}{2}bh$.

   Remark: If b is increased by c times itself (to $b + cb$), and h is decreased by c times itself (to $h - ch$), then their product $p = bh$ is decreased by c^2 times itself, to

$$p' = (1 + c)b(1 - c)h = (1 - c^2)bh = (1 - c^2)p = p - c^2p.$$

3. (D) Let r and s denote the two numbers with given arithmetic and geometric means:

$$\tfrac{1}{2}(r + s) = 6, \qquad \sqrt{rs} = 10.$$

   Then $r + s = 12$ and $rs = 100$. A quadratic equation with roots r and s is

$$(x - r)(x - s) = x^2 - (r + s)x + rs = 0;$$

   and, when the above values for $r + s$ and rs are used, the equation $x^2 - 12x + 100 = 0$, given in (D), is obtained.

---

† The letter following the problem number refers to the correct choice of the five listed in the examination.

Remark: It is clear that r and s cannot be positive real numbers; for, if they were, the given data would contradict the famous arithmetic-geometric mean inequality which states that, for positive real numbers a and b,

$$\frac{a+b}{2} \geq \sqrt{ab};$$

the equality holds if and only if $a = b$.† Indeed, the required equation in (D) has the conjugate complex roots $6 \pm 8i$.

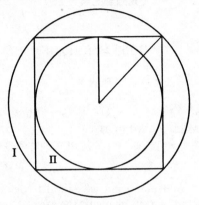

4. (B) The ratio of the radius of the circumscribed circle to that of the inscribed circle is the same as the ratio of the diagonal of the square to its side, and this is $\sqrt{2}$. The ratio of the areas of the circles is the square of the ratio of their radii; and $(\sqrt{2})^2 = 2$.

5. (A) The left member of the given equation is not defined when $x = 0$, nor when $x = 5$; for all other x it has the constant value 2. The right member is defined for all x and has the value 2 only when $x = 5$, so that no value of x satisfies the given equation.

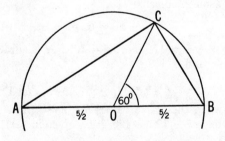

† For a full discussion of this inequality and its generalizations, see NML 12, pp. 70–72.

6. (C) Triangle ABC is a right triangle with hypotenuse $AB = 5$ inches because $\measuredangle ACB$ is inscribed in a semicircle. Since radii $OC = OB = \frac{5}{2}$ inches, and angle BOC between them is 60°, triangle BOC is equilateral, and leg BC of right triangle ABC is $\frac{5}{2}$ inches. The Pythagorean theorem now yields

$$AC^2 = AB^2 - BC^2 = 5^2 - \left(\frac{5}{2}\right)^2 = \frac{3}{4}(25), \quad AC = \frac{5\sqrt{3}}{2}.$$

7. (A) When the right side of the given identity is written as a single fraction, the identity reads

$$\frac{35x - 29}{x^2 - 3x + 2} = \frac{(x - 2)N_1 + (x - 1)N_2}{(x - 1)(x - 2)}.$$

The denominators are identical, hence the numerators are also identical. This means

$$35x - 29 = (x - 2)N_1 + (x - 1)N_2 \qquad \text{for all } x.$$

A linear function of x is completely determined by its values at two distinct points x_1 and x_2 which, in the above case, are most conveniently taken to be $x_1 = 1$ and $x_2 = 2$. Substitution of these values for x yields

$$35 - 29 = -N_1, \qquad N_1 = -6,$$

and

$$70 - 29 = N_2, \qquad N_2 = 41,$$

respectively. Thus $N_1 N_2 = -246$.

Remark: Since two linear functions of x are the same if and only if the coefficients of x are equal and their constant terms are equal, this problem can also be solved by equating these corresponding coefficients. This leads to two linear equations in N_1 and N_2 with solution $N_1 = -6$, $N_2 = 41$.

The argument in the above solution as well as that outlined in this remark can be generalized to polynomials of higher degrees.

8. (B) Denote the common chord by AB, its midpoint by P, and the centers of the smaller and larger circles by O and O'; OO' is perpendicular to AB and passes through P. The Pythagorean theorem applied to right triangles OPA and $O'PA$ now yields

$$OP^2 = OA^2 - AP^2 = 10^2 - 8^2 = 36, \qquad OP = 6,$$

and

$$PO'^2 = O'A^2 - AP^2 = 17^2 - 8^2 = 225, \quad PO' = 15.$$

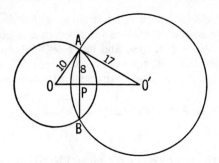

If, as in the figure, O lies outside the larger circle, the distance between the centers is

$$OO' = OP + PO' = 6 + 15 = 21.$$

(If O lies inside the larger circle, then $OO' = PO' - OP = 15 - 6 = 9$.)

9. (A) Since, by the definition of logarithms,

$$\log_8 2 = \tfrac{1}{3} \quad \text{and} \quad \log_2 8 = 3,$$

we have

$$x = (\log_8 2)^{(\log_2 8)} = (\tfrac{1}{3})^3 = 3^{-3},$$

so that $\log_3 x = -3$.

10. (E) Let the numbers be x and y. Then since

(1) $$x + y = xy = 1,$$

$$1 = (x + y)^3 = x^3 + y^3 + 3xy(x + y) = x^3 + y^3 + 3(1)(1)$$
$$= x^3 + y^3 + 3.$$
$$\therefore \ x^3 + y^3 = -2.$$

OR

The numbers x, y satisfy the quadratic equation

(2) $$u^2 - u + 1 = 0$$

since their sum is 1 and their product is 1. [Eq. (2) can also be derived from (1) by substitution of $1 - x$ for y in $xy = 1$.] We observe that

(3) $$(u + 1)(u^2 - u + 1) = u^3 + 1 = 0,$$

so that the roots of (2) also satisfy (3); hence they are cube roots of -1. Thus we have $x^3 = -1$ and $y^3 = -1$, whence $x^3 + y^3 = -2$.

If one does not recognize $u^2 - u + 1$ as a factor of $u^3 + 1$, one can compute the roots $x = \frac{1}{2}(1 + \sqrt{3}i)$, $y = \frac{1}{2}(1 - \sqrt{3}i)$ explicitly, cube each, and add the result.

11. (C) Since an angle bisector of a triangle divides the opposite side into segments proportional to the adjacent sides, we have the shortest side AC of length 10 divided by D in the ratio 4:3. Thus the longer segment is $\frac{4}{7}$ of the length of AC; that is, $\frac{4}{7} \cdot 10 = 5\frac{5}{7}$ is the length of the longer segment.

12. (E) In terms of only powers of 2, the given equation is equivalent to

$$(2^{6x+3})(2^{2(3x+6)}) = 2^{3(4x+5)} \quad \text{or} \quad 2^{12x+15} = 2^{12x+15},$$

which is true for all real values of x.

13. (E) Between every pair of distinct real numbers, there are infinitely many rational numbers. In particular, between 0 and 5 there are infinitely many rational numbers x such that $0 < x < 5$. Pick any such x and set $y = 5 - x$; then $0 < y < 5$. Moreover, y is rational, and $x + y = 5$ (so that in particular $x + y \leq 5$).

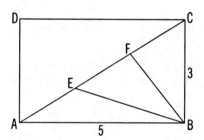

14. (C) Triangles AEB, BEF, and FCB have equal areas because they have the same altitude from B and equal bases. Hence each has area one-third the area of triangle ABC, that is $\frac{1}{3}(\frac{1}{2} \cdot 5 \cdot 3) = \frac{5}{2}$ square inches.

15. (D) Since $x - y > x$, $-y > 0$ and $y < 0$.

Since $x + y < y$, $x < 0$. $\therefore x < 0$, $y < 0$.

16. (B) In terms of powers of 2, the first equation gives

$$\frac{2^{2x}}{2^{x+y}} = 2^{x-y} = 2^3. \quad \therefore x - y = 3.$$

In terms of powers of 3, the second equation gives

$$\frac{3^{2(x+y)}}{3^{5y}} = 3^{2x-3y} = 3^5. \quad \therefore \quad 2x - 3y = 5.$$

The equations $x - y = 3$ and $2x - 3y = 5$ now yield $y = 1$, $x = 4$, so that $xy = 4$.

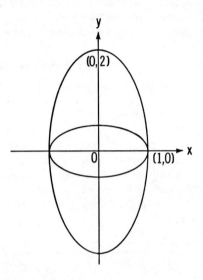

17. (C) Both curves are ellipses with centers at the origin and axes along the coordinate axes. In standard form, their equations are

$$\frac{x^2}{1^2} + \frac{y^2}{(1/2)^2} = 1 \quad \text{and} \quad \frac{x^2}{1^2} + \frac{y^2}{2^2} = 1,$$

from which we see that the first has, as its major axis, the segment from $(-1, 0)$ to $(1, 0)$, while the second has that same segment for its minor axis, so that all other points of the second ellipse lie outside the first. We conclude that the only points common to both curves are the points $(-1, 0)$ and $(1, 0)$ where the ellipses are tangent.

OR

The coordinates (x, y) of a point common to both curves satisfy both equations, hence also their sum $5x^2 + 5y^2 = 5$, which describes the unit circle with center at the origin. The only points on both ellipses and on this circle are the points $(1, 0)$ and $(-1, 0)$ where both ellipses are tangent to the circle.

18. (A) Formulas for the n-th term l and the sum s of n terms of an A.P. with first term a and common difference d are (see p. 114)
$$l = a + (n - 1)d \quad \text{and} \quad s = \tfrac{1}{2}n(a + l).$$
The second formula gives $155 = \tfrac{1}{2}n(2 + 29)$, $n = 10$, and the first now yields $29 = 2 + (10 - 1)d$, $d = 3$.

19. (B) The formula for the sum s of the first n terms of an A.P. may be written as $s = \dfrac{n}{2}[2a + (n - 1)d]$, see p. 114. Hence

$$s_1 = \frac{n}{2}[2\cdot 8 + (n - 1)4] = \frac{n}{2}[12 + 4n]$$

$$s_2 = \frac{n}{2}[2\cdot 17 + (n - 1)2] = \frac{n}{2}[32 + 2n]$$

and, for $n \neq 0$, $s_1 = s_2$ if and only if the expressions in brackets are equal, i.e., if and only if $12 + 4n = 32 + 2n$ or $n = 10$.

20. (C) Let $P(a, b)$ be a proposition concerning a and b. Then the the negation of the statement "For all a and b, $P(a, b)$ holds" is "there exist a and b such that $P(a, b)$ does not hold." In the present case $P(a, b)$ is the statement: if $a = 0$ and b is any real number, then $ab = 0$. This is equivalent to saying that "either $a \neq 0$ or $ab = 0$." Now the denial of a statement of the form "either S or T" is "not S and not T." Hence the denial of the statement $P(a, b)$ is "$a = 0$ and $ab \neq 0$." Thus, the negation of the statement "For all a and b, $P(a, b)$ holds" is "there exist a and b such that $a = 0$ and $ab \neq 0$."

Part 2

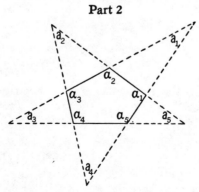

21. (E) Denote the angles of the star in the figure by a_1, a_2, a_3, $\cdots$, a_n,

and the angles of the small convex polygon by $\alpha_1, \alpha_2, \alpha_3, \cdots,$ α_n with $\alpha_{n+1} = \alpha_1$. Then

$$a_1 = 180 - (180 - \alpha_1) - (180 - \alpha_2) \quad = \alpha_1 + \alpha_2 - 180,$$

$$a_2 = 180 - (180 - \alpha_2) - (180 - \alpha_3) \quad = \alpha_2 + \alpha_3 - 180,$$

$$\cdots \qquad \cdots \qquad \cdots \qquad \cdots \qquad \cdots \qquad\qquad \cdots \qquad \cdots$$

$$a_n = 180 - (180 - \alpha_n) - (180 - \alpha_{n+1}) = \alpha_n + \alpha_1 - 180.$$

Summing the angles on the left and those on the far right, we get

$$S = 2(\alpha_1 + \alpha_2 + \alpha_3 + \cdots + \alpha_n) - 180n.$$

Since the sum of the n interior angles of a convex n-sided polygon is $180(n - 2)$,

$$S = 2 \cdot 180(n - 2) - 180n = 180(n - 4) \text{ degrees.}$$

22. (A) Each of the four equations has an infinite number of solutions $(a, b) \neq (0, 0)$. For example, if a is any number greater than 1, then I, II, III and IV are satisfied when b is equal to $a\sqrt{-1}$, $a/\sqrt{a^2 - 1}$, 0, and 0, respectively. There are many other solutions of each equation.

23. (A) We may treat the equation as a quadratic in y with discriminant

$$D = (4x)^2 - 4 \cdot 4(x + 6) = 16(x^2 - x - 6)$$

$$= 16(x - 3)(x + 2).$$

Now y is real if and only if $D \geq 0$, and this is true when both factors on the right, $x - 3$ and $x + 2$, have like signs, i.e., when $x \leq -2$ or $x \geq 3$. Alternatively, the equation may be written

$$4y^2 + 4xy + x^2 - (x^2 - x - 6) = 0$$

or

$$(2y + x)^2 = (x + 2)(x - 3).$$

For real x, the left member of the last equation is non-negative if and only if y is real, so the product $(x + 2)(x - 3) \geq 0$, which is true when $x \leq -2$ or $x \geq 3$.

24. (B) The identity $(\log_N M)(\log_M N) = 1$ together with the given equation yields $(\log_N M)^2 = 1$. $\therefore \log_N M = 1$ or -1. If $\log_N M = 1$, then $M = N$ which is ruled out. We conclude that $\log_N M = -1$. $\therefore M = N^{-1}$, $MN = 1$.

OR

If $\log_N M = x = \log_M N$, then $M = N^x$, and $N = M^x = (N^x)^x = N^{x^2}$. Since $N \neq 1$, we conclude $x^2 = 1$, so $x = 1$ or $x = -1$. We reject $x = 1$ (since it leads to $M = N$) and conclude $x = -1$, so $MN = 1$.

25. (D) Since $F(n+1) = F(n) + \frac{1}{2}$, the sequence $F(n)$ is an arithmetic progression with first term $F(1) = 2$ and common difference $\frac{1}{2}$. The 101st term is

$$F(101) = 2 + (101 - 1)\tfrac{1}{2} = 2 + 50 = 52.$$

26. (C) Substituting y from the second equation into the first gives $13x + 11(mx - 1) = 700$, so that

$$x = \frac{711}{13 + 11m} = \frac{3^2 \cdot 79}{13 + 11m}.$$

Since x is to be an integer, the denominator $13 + 11m$ must be a divisor of the numerator, and its only divisors are $1, 3, 3^2, 79, 3 \cdot 79, 3^2 \cdot 79$. Our task now is to find a positive integer m such that

$$13 + 11m = d, \quad \text{or} \quad m = \frac{d - 13}{11},$$

where d is one of these divisors. Since $m > 0$, we see that $d > 13$, so the only divisors we need to test are the last three:

(i)   if $d = 79$, $d - 13 = 66$, and $m = \frac{66}{11} = 6$

(ii) if $d = 3 \cdot 79 = 237$, $d - 13 = 224$ is not divisible by 11

(iii) if $d = 3^2 \cdot 79 = 711$, $d - 13 = 698$ is not divisible by 11.

We conclude that $m = 6$ is the only positive integer yielding a lattice point for the intersection of the given lines.

27. (A) Let c denote the speed of the current and m the usual rate of the rower in still water. Then his downstream and upstream rates are $m + c$ and $m - c$, respectively, and after he doubles his usual rate, they are $2m + c$ and $2m - c$. Since the distance is 15 miles and the time is distance/rate, the problem tells us that

$$\frac{15}{m + c} = \frac{15}{m - c} - 5 \quad \text{and} \quad \frac{15}{2m + c} = \frac{15}{2m - c} - 1.$$

Multiplying the first equation by $(m+c)(m-c)$ and the second by $(2m+c)(2m-c)$ and simplifying yields

$$5m^2 - 5c^2 = 30c \quad \text{and} \quad 4m^2 - c^2 = 30c,$$

and subtracting the second from the first gives

$$m^2 - 4c^2 = 0, \quad \text{so} \quad m = 2c.$$

If we substitute for m, in the second equation above, we get

$$4(4c^2) - c^2 = 30c \quad \text{so} \quad 15c = 30, \quad c = 2.$$

28. (B) The proportion $AP:PD = BP:PC$ may be written as

$$(p-a):(d-p) = (p-b):(c-p),$$

where p denotes the distance OP.

$$\therefore \ (-a+p)(-p+c) = (-b+p)(-p+d),$$

$$-ac + (a+c)p - p^2 = -p^2 + (b+d)p - bd,$$

$$[(a+c) - (b+d)]p = ac - bd,$$

$$p = \frac{ac - bd}{a - b + c - d} = OP.$$

29. (B) The number of positive integers less than some positive integer M is $M - 1$; the number of positive integers less than M and divisible by d is the greatest integer not exceeding $(M - 1)/d$. (We denote the greatest integer not exceeding the number x by the symbol $[x]$.) Thus, among the 999 positive integers less than 1000, there are $N_1 = [999/5]$ which are divisible by 5 and $N_2 = [999/7]$ which are divisible by 7. But some of the numbers divisible by 5 are also divisible by 7. Therefore, if we subtract from the 999 numbers less than 1000 all the numbers divisible by 5, and subsequently all the numbers divisible by 7, we would be subtracting those divisible by 5 and 7 (i.e., by 35) twice, once in each batch. Therefore, the desired answer is

$$999 - \frac{999}{5} - \frac{999}{7} + \frac{999}{35} = 999 - 199 - 142 + 28$$

$$= 686.$$

30. (D) Since the sum of the roots is zero, the 4th root is -6 and the equation is $(x - 2)(x - 3)(x - 1)(x + 6) = 0$.

$$\therefore \ (x^2 - 5x + 6)(x^2 + 5x - 6) = x^4 - (5x - 6)^2$$

$$= x^4 - 25x^2 + 60x - 36 = 0.$$

$$\therefore a + c = -25 - 36 = -61.$$

OR

We note that a is the sum of the products of the roots taken two at a time, and c is the product of the four roots. Hence

$$a = 1 \cdot 2 + 1 \cdot 3 + 1 \cdot (-6) + 2 \cdot 3 + 2 \cdot (-6) + 3 \cdot (-6)$$

$$= -25$$

and $c = 1 \cdot 2 \cdot 3 \cdot (-6) = -36.$ $\therefore a + c = -61.$

OR

Substituting the three roots into the given equation gives three equations in the three unknowns a, b, and c; viz.

$$1 + a + b + c = 0$$

$$16 + 4a + 2b + c = 0$$

$$81 + 8a + 3b + c = 0$$

with solution $(a, b, c) = (-25, 60, -36)$ from which

$$a + c = -25 - 36 = -61.$$

Part 3

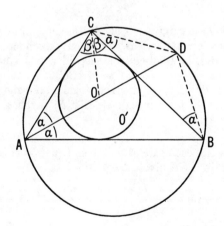

31. (D) Since AB and AC are tangent to the small circle and AD passes through its center, we have $\angle CAD = \angle BAD = \alpha$ (see figure). Similarly, $\angle ACO = \angle BCO = \beta$. Therefore arcs CD and BD are equal and hence also the chords: $CD = BD$.

Next we show that $CD = OD$ by proving that these are sides opposite equal angles in $\triangle CDO$, i.e. by proving $\angle OCD = \angle COD$. Now $\angle OCD = \angle OCB + \angle BCD = \beta + \alpha$, and $\angle COD$ is an exterior angle of $\triangle AOC$, hence equal to the sum $\alpha + \beta$ of the remote interior angles. We conclude that $CD = OD$, and that (D) is the correct answer.

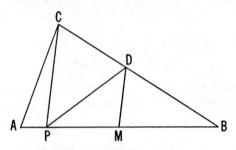

32. (B) Denote the area of a triangle XYZ by (XYZ). Since $PC \parallel MD$, $(MDC) = (MDP)$. It follows that

$$(BPD) = (BMD) + (MDP) = (BMD) + (MDC)$$

$$= (BMC) = \tfrac{1}{2}(ABC),$$

since CM is a median. Thus

$$r = \frac{(BPD)}{(ABC)} = \frac{1}{2}.$$

Remark: If P lies to the left of A, the same proof works; but if P lies between M and B, PC is the common base of triangles PCM and PCD of equal areas which, when added to $\triangle PCB$, yield $(BPD) = (BMC) = \tfrac{1}{2}(ABC)$. What if P lies to the right of B? Can you devise a single proof valid for any position of P on the line through A and B, perhaps with the help of signed areas?

33. (D) If we write each side of the given equation as a single fraction, we obtain the equivalent equation

$$\frac{a(x - a) + b(x - b)}{ab} = \frac{b(x - b) + a(x - a)}{(x - a)(x - b)}.$$

We observe that the numerators are identical. Thus the equation is satisfied either (i) if the denominators are equal, or (ii) if the numerator is zero.

Case (i) requires

$$(x - a)(x - b) = x^2 - (a + b)x + ab = ab,$$

so $x[x - (a + b)] = 0$, and this is true if $x = 0$ or $x = a + b$. Case (ii) requires that

$$a(x - a) + b(x - b) = (a + b)x - (a^2 + b^2) = 0,$$

and this is true if $x = (a^2 + b^2)/(a + b)$. The given conditions on a and b ensure that the three values we have found are indeed distinct.

34. (B) We are told that $r = \text{speed (in m/hr)} = \dfrac{\text{distance (in miles)}}{\text{time (in hours)}}$.

Then

$$r \cdot \frac{5280}{3600} = \frac{22}{15} r = \frac{\text{distance (in feet)}}{\text{time (in seconds)}},$$

and

$$\text{time (in seconds)} = \frac{\text{distance (in feet)}}{r} \cdot \frac{15}{22}.$$

Time for one rotation $= \dfrac{11 \cdot 15}{22r} = \dfrac{15}{2r} = t$ seconds. When r is increased by 5, t is decreased by $\frac{1}{4}$, so

$$\frac{15}{2(r + 5)} = t - \frac{1}{4}.$$

It follows that

$$\frac{15}{2(r + 5)} = \frac{15}{2r} - \frac{1}{4} = \frac{30 - r}{4r},$$

so that $30r = (r + 5)(30 - r) = 30r - r^2 + 150 - 5r$, or $r^2 + 5r - 150 = 0$, $(r - 10)(r + 15) = 0$. We reject the negative speed and conclude that $r = 10$.

35. (C) The first member of each of the following inequalities compares the base of a triangle with the sum of the other two sides. The second member (which we prove below) compares that sum with the sum of the sides of another triangle having the same base but containing the first triangle, see figure on next page. Thus

$$AB < OA + OB < AC + CB$$
$$BC < OB + OC < BA + AC$$
$$CA < OC + OA < CB + BA.$$

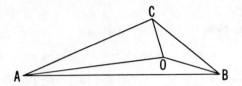

Adding these member by member, we get

$$s_2 < 2s_1 < 2s_2, \qquad \tfrac{1}{2}s_2 < s_1 < s_2$$

as required for every triangle in choice (C).

To see that $OA + OB < AC + CB$, extend segment AO to meet side BC in point D. By the triangle inequality,

$$AD = AO + OD < AC + CD,$$

$$OB < OD + DB,$$

and addition yields

$$AO + OD + OB < AC + CD + OD + DB,$$

so

$$AO + OB < AC + CD + DB = AC + CB.$$

36. (E) Letting first $x = 1$ and then $x = -1$ in the given identity, we obtain

$$(1 + 1 + 1^2)^n = 3^n = a_0 + a_1 + a_2 + \cdots + a_{2n-1} + a_{2n}$$

$$[1 + (-1) + (-1)^2]^n = 1^n$$

$$= a_0 - a_1 + a_2 - \cdots - a_{2n-1} + a_{2n}.$$

Adding these, we get

$$3^n + 1 = 2a_0 + 2a_2 + \cdots + 2a_{2n}$$

$$= 2(a_0 + a_2 + \cdots + a_{2n}) = 2s,$$

$$s = \frac{3^n + 1}{2}.$$

Remark: When $n = 1$, we get $s = a_0 + a_2 = 1 + 1 = 2$. This eliminates choices (B), (C), and (D). In a similar way, one can eliminate choice (A) by taking $n = 2$.

37. (C) Let a, b and c denote the number of hours it takes Alpha, Beta, and Gamma, respectively, to complete the job alone. Then the fraction of the job each completes in one hour is $1/a$, $1/b$, $1/c$, respectively, and the fraction of the job all three together

complete in one hour is $1/a + 1/b + 1/c$. Moreover, needing "6 hours less time than Alpha alone," "one hour less than Beta" and "half the time needed by Gamma" leads to hourly rates of

$$\frac{1}{a-6}, \frac{1}{b-1}, \text{ and } \frac{1}{c/2} = \frac{2}{c}, \text{ respectively. So the given in-}$$

formation may be written in the form

$$\frac{1}{a} + \frac{1}{b} + \frac{1}{c} = \frac{1}{a-6} = \frac{1}{b-1} = \frac{2}{c},$$

and

$$\frac{1}{a} + \frac{1}{b} = \frac{1}{h} \text{ so that } \frac{1}{h} + \frac{1}{c} = \frac{2}{c}, \quad \frac{1}{h} = \frac{1}{c}, \quad h = c.$$

From $\dfrac{1}{a-6} = \dfrac{2}{h}$ we obtain $a = \dfrac{h+12}{2}$; from $\dfrac{1}{b-1} = \dfrac{2}{h}$,

we get $b = \dfrac{h+2}{2}$. The sum of the reciprocals of a and b

is the reciprocal of h, so that $\dfrac{2}{h+12} + \dfrac{2}{h+2} = \dfrac{1}{h}$. Clearing

fractions and simplifying leads to

$$3h^2 + 14h - 24 = 0 \quad \text{or} \quad (3h-4)(h+6) = 0.$$

The negative value of h is ruled out, so $h = \frac{4}{3}$.

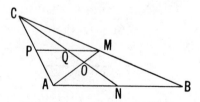

38. (D) The base OQ of $\triangle OMQ$ is equal to

$$OQ = CO - CQ = \tfrac{2}{3}CN - \tfrac{1}{2}CN = \tfrac{1}{6}CN.$$

Let h be the altitude of $\triangle OMQ$ from M to side OQ. then $2h$ is the altitude from B of $\triangle CNB$. Thus

$$\text{Area of } \triangle OMQ = \tfrac{1}{2}OQ \cdot h = \tfrac{1}{12}CN \cdot h = n.$$

$$\text{Area of } \triangle ABC = 2 \ (\text{Area of } \triangle CNB)$$

$$= 2(\tfrac{1}{2}CN \cdot 2h) = 2CN \cdot h = 24n.$$

39. (E) We write each fraction as an infinite series using $1/R_1$ then $1/R_2$ as common ratio; we sum the infinite series and equate corresponding expressions for F_1 and F_2:

$$F_1 = \frac{3R_1 + 7}{R_1^2} + \frac{3R_1 + 7}{R_1^4} + \cdots = \frac{3R_1 + 7}{R_1^2 - 1}$$

$$= \frac{2R_2 + 5}{R_2^2} + \frac{2R_2 + 5}{R_2^4} + \cdots = \frac{2R_2 + 5}{R_2^2 - 1}.$$

$$\therefore F_1 = \frac{3R_1 + 7}{R_1^2 - 1} = \frac{2R_2 + 5}{R_2^2 - 1};$$

similarly,

$$F_2 = \frac{7R_1 + 3}{R_1^2 - 1} = \frac{5R_2 + 2}{R_2^2 - 1}.$$

Now

$$F_1 + F_2 = \frac{10R_1 + 10}{(R_1 + 1)(R_1 - 1)} = \frac{10}{R_1 - 1}$$

$$= \frac{7R_2 + 7}{(R_2 + 1)(R_2 - 1)} = \frac{7}{R_2 - 1};$$

$$\therefore \frac{R_1 - 1}{10} = \frac{R_2 - 1}{7}, \qquad 7R_1 - 10R_2 + 3 = 0.$$

$$F_2 - F_1 = \frac{4R_1 - 4}{(R_1 + 1)(R_1 - 1)} = \frac{4}{R_1 + 1}$$

$$= \frac{3R_2 - 3}{(R_2 + 1)(R_2 - 1)} = \frac{3}{R_2 + 1}$$

$$\therefore \frac{R_1 + 1}{4} = \frac{R_2 + 1}{3}, \qquad 3R_1 - 4R_2 - 1 = 0.$$

The two linear equations in R_1 and R_2 have the unique solution $R_1 = 11$, $R_2 = 8$, whence $R_1 + R_2 = 19$.†

---

† Editor's note: The reduction to a linear system of equations was due to the fact that F_1 and F_2 have the same digits in reverse order. For arbitrary F_1 and F_2 with period 2, we must expect a quadratic system of equations.

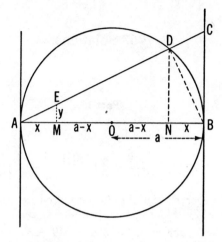

40. (A) Let perpendiculars from points E and D meet diameter AB at M and N respectively. Since transversal AC intercepts equal segments $(AE$ and $DC)$ between parallel lines, transversal AB does likewise, so $NB = x$. The altitude ND to the hypotenuse of right triangle ABD is the mean proportional between the segments $AN = (2a - x)$ and $NB = x$, i.e. $ND^2 = x(2a - x)$. Similar right triangles AME and AND give

$$\frac{ND}{2a - x} = \frac{y}{x}, \qquad ND = \frac{y(2a - x)}{x}$$

$$\therefore \; \frac{y^2(2a - x)^2}{x^2} = (2a - x) \cdot x \quad \text{or} \quad y^2 = \frac{x^3}{2a - x}.$$

1967 Solutions

Part 1

1. **(C)** Since $5b9$ is divisible by 9, and $0 \le b \le 9$,

$$\frac{5b9}{9} = 10\frac{(50 + b)}{9} + 1$$

is an integer; so $(50 + b)/9$ must also be an integer. $\therefore$ **$b = 4$**. Now

$$2a3 = 5b9 - 326 = 549 - 326 - 223.$$

Therefore $a = 2$, and $a + b = 2 + 4 = 6$.

OR

One may note directly that the sum $(5 + b + 9)$ of the digits in $5b9$ must be a multiple of 9. (A number is divisible by 9 if and only if the sum of its digits is divisible by 9.) So the digit b is 4. We have then, as before, $a = 2$ and $a + b = 6$.

OR

Using congruences, $5b9$ is congruent to 0 modulo 9,

$$500 + 10b + 9 \equiv 5 + b \equiv 0 \bmod 9, \quad b \equiv -5 \equiv 4 \bmod 9.$$

$\therefore$ $b = 4$ because $0 \le b \le 9$. Then $a = 2$, $a + b = 6$ follow as in the other solutions.

2. **(D)** The given expression is equivalent to

$$\left(x + \frac{1}{x}\right)\left(y + \frac{1}{y}\right) + \left(x - \frac{1}{x}\right)\left(y - \frac{1}{y}\right)$$

$$= \left(xy + \frac{y}{x} + \frac{x}{y} + \frac{1}{xy}\right) + \left(xy - \frac{y}{x} - \frac{x}{y} + \frac{1}{xy}\right)$$

$$= 2xy + \frac{2}{xy}.$$

OR

Performing the multiplications in each of its two terms, the given expression can be written as

$$\frac{x^2y^2 + x^2 + y^2 + 1}{xy} + \frac{x^2y^2 - x^2 - y^2 + 1}{xy} = \frac{2x^2y^2 + 2}{xy}$$

$$= 2xy + \frac{2}{xy}$$

as before.

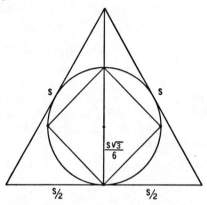

3. (B) The altitude h of the given triangle is $s\sqrt{3}/2$, and the radius r of the inscribed circle is

$$r = \frac{h}{3} = \frac{s\sqrt{3}}{6}.$$

The diagonal of the inscribed square is the diameter $2r = s\sqrt{3}/3$ of the circle, and the area of the square is half the product of its diagonals:

$$\text{area of square} = \frac{(2r)^2}{2} = \frac{s^2 3}{2 \cdot 9} = \frac{s^2}{6}.$$

OR

We could have found the side of the square first by dividing its diagonal by $\sqrt{2}$, then computed the area of the square by squaring its side:

$$\left(\frac{s\sqrt{3}}{3\sqrt{2}}\right)^2 = \frac{3s^2}{18} = \frac{s^2}{6}.$$

4. (C) The first three given logarithmic equalities are equivalent to the exponential equalities $a = x^p$, $b = x^q$, $c = x^r$ whence

$$\frac{b^2}{ac} = \frac{x^{2q}}{x^{p+r}} = x^{2q-p-r} = x^y, \qquad y = 2q - p - r.$$

Alternatively, we may express the relation involving y in logarithmic form:

$$y \log x = 2 \log b - \log a - \log c.$$

Substituting for the logarithms on the right from the first given relations yields

$$y \log x = 2q \log x - p \log x - r \log x,$$

and, since $x \neq 1$, $\log x \neq 0$, so division by $\log x$ yields

$$y = 2q - p - r.$$

5. (D) Denote the sides of the circumscribed triangle by a, b, and c (see figure). Then the radii to the points of contact are perpendicular to the sides and hence are altitudes of the three triangles into which the circumscribed triangle is partitioned. Its area is therefore

$$K = (ar + br + cr)/2 = (a + b + c)r/2 = Pr/2$$

so that $P/K = 2/r$.

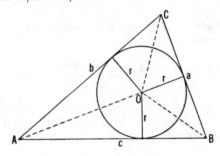

Remark: Clearly P is proportional to r, and K is proportional to r^2, so P/K is proportional to $1/r$. This eliminates choices (A), (C), and (E).

6. (D) From $f(x) = 4^x$, we obtain

$$f(x + 1) - f(x) = 4^{x+1} - 4^x$$
$$= 4 \cdot 4^x - 4^x = (4 - 1)4^x = 3 \cdot 4^x = 3f(x).$$

7. (E) When b, $(-c)$, and d are all positive, the given inequality is equivalent to $a < -bc/d$, which means that a is less than the positive number $-bc/d$. This will be true when a is positive but less than $-bc/d$, or when $a = 0$, or when a is negative.

8. (A) The final mixture of $m + x$ ounces contains $(m/100)m$ ounces of acid, and this is to be $(m - 10)/100$ of the mixture. So

$$\frac{m - 10}{100}(m + x) = \frac{m^2}{100}.$$

When solved for x, this yields

$$x = \frac{10m}{m - 10}.$$

[If m were less than 10 ounces, water would have to be extracted from, rather than added to, the initial mixture of m ounces.]

9. (E) Let the shorter base, altitude, and longer base be denoted by $(a - d)$, a, and $(a + d)$, respectively. Then the area is

$$K = \tfrac{1}{2}a(a - d + a + d) = a^2.$$

Since we have no knowledge about the nature of the number a, we cannot deduce properties of a^2; so (E) is the correct answer.

10. (A) Multiplying both members of the given identity by the positive number $(10^x - 1)(10^x + 2)$ gives

$$a(10^x + 2) + b(10^x - 1) = 2 \cdot 10^x + 3.$$

Equating the constant terms and the coefficients of 10^x in the two members of this identity yields $2a - b = 3$ and $a + b = 2$, respectively. Solving this system of linear equations in a and b gives $3a = 5$, $a = \tfrac{5}{3}$, so that $\tfrac{5}{3} + b = 2$, $b = \tfrac{1}{3}$. Hence $a - b = \tfrac{4}{3}$ as stated in choice (A).

11. (B) If we denote the length of side AB by x, then the length of the adjacent side BC is $(10 - x)$. Sides AB and BC of rectangle $ABCD$ are legs of the right triangle ABC whose hypotenuse is the diagonal AC. The Pythagorean Theorem now yields

$$AC^2 = AB^2 + BC^2 = x^2 + (10 - x)^2$$
$$= 2(x^2 - 10x + 50) = 2[(x - 5)^2 + 25],$$

which takes on its least value, 50, when $(x - 5)^2 = 0$, i.e., $x = AB = 5$ and $(10 - x) = BC = 5$. Thus the rectangle of perimeter 20 with least diagonal is a square, and the length of diagonal AC is $\sqrt{50} = 5\sqrt{2}$ inches.

Remark: Denote the lengths of AB, BC and AC by the positive numbers a, b, and c respectively. Then the perimeter $p = 2(a+b)$, and minimizing c is equivalent to minimizing

$$c^2 = a^2 + b^2 = (a+b)^2 - 2ab = \frac{p^2}{4} - 2ab.$$

If p is fixed, c^2 is smallest when ab is largest. We apply the arithmetic-geometric mean inequality

$$\frac{p}{4} = \frac{a+b}{2} \geq \sqrt{ab},$$

where $=$ holds if and only if $a = b$; since the left member is constant, $\sqrt{ab}$ (and hence ab) is largest when $a = b$. Thus the rectangle is a square, and this argument holds for any given perimeter.

12. (B) The area described (left figure) is that of a trapezoid with bases of lengths $(m+4)$ along $x = 1$ and $(4m+4)$ along $x = 4$, and altitude of length 3 along the x-axis. So

$$\frac{1}{2}(3)(m+4+4m+4) = 7, \qquad 5m = \frac{-10}{3}, \qquad m = -\frac{2}{3}.$$

[The convexity of the region rules out the possibility of the line $y = mx + 4$ crossing the x-axis in the interval $1 < x < 4$.]

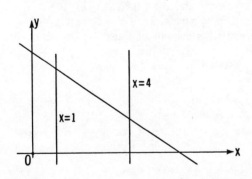

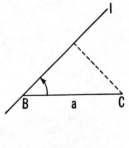

13. (E) We begin the construction with the given side a and denote its endpoints by B and C (right figure). At B we construct a line l making the given angle B with side a. Now either

    (i) the distance from C to l is h_c, in which case vertex A of $\triangle ABC$ may be placed anywhere on line l (infinitely many solutions), or

(ii) the distance from C to l is not h_c, in which case no triangle satisfies the given conditions (no solution).

14. (C) Since $f(t) = t/(1 - t)$, $t \neq 1$, $y = f(x)$ is equivalent to $y = x/(1 - x)$ which may be solved for x:

$$x = \frac{y}{1 + y} = -\frac{(-y)}{1 - (-y)} = -f(-y).$$

15. (D) Let T $(T + 18)$ denote the area in square feet of the smaller (larger) triangle, and 3 and x denote the corresponding sides in feet. Since the areas of similar triangles are proportional to the squares of corresponding sides,

$$\frac{T + 18}{T} = \frac{x^2}{3^2} = \left(\frac{x}{3}\right)^2,$$

By assumption, $x/3$ is an integer, so x is a multiple of 3. Solving this equation for T yields

$$T = \frac{18}{(x/3)^2 - 1},$$

and since T is required to be an integer, $(x/3)^2 - 1$ is a divisor of 18. Thus $(x/3)^2 = 2, 3, 4, 7, 10,$ or 19. The only square among these numbers is 4; hence $(x/3)^2 = 4$, $x/3 = 2$, and $x = 6$.

16. (B) The given equality $(12)(15)(16) = 3146$ in base b means

$$(b + 2)(b + 5)(b + 6) = 3b^3 + b^2 + 4b + 6.$$

After some simplification, we obtain the equivalent equation

$$b^3 - 6b^2 - 24b - 27 = 0.$$

Its only real solution is $b = 9$. The sum $s = 12 + 15 + 16$ in base b means

$$s = (b + 2) + (b + 5) + (b + 6)$$
$$= 3b + 13 = 3b + b + 4 = 4b + 4$$

which, in base $b = 9$, is written 44.

17. (A) Since the roots of the given quadratic equation are real and distinct, its discriminant $(p^2 - 32)$ is positive. Hence $p^2 > 32$, $|p| > 4\sqrt{2}$. But the sum of the roots is $r_1 + r_2 = -p$, and hence $|r_1 + r_2| = |-p| = |p| > 4\sqrt{2}$.

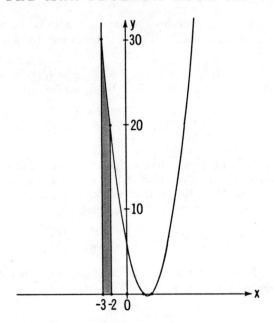

18. (B) Set $x^2 - 5x + 6 = F(x)$; its factored form is $F(x) = (x - 2)(x - 3)$, and the condition that $F(x) < 0$ implies that the two factors have unlike signs, which occurs only when $2 < x < 3$ (see figure). The second given function $P(x) = x^2 + 5x + 6$ increases as x increases from 2 to 3. Its least value is $P(2) = 4 + 10 + 6 = 20$, its greatest value is $P(3) = 9 + 15 + 6 = 30$, and it takes on all values between 20 and 30 as x varies from 2 to 3.

19. (E) Denote the length and width in inches of the rectangle by l and w respectively; then its area lw satisfies

$$lw = (l + \tfrac{5}{2})(w - \tfrac{2}{3}), \quad \text{and} \quad lw = (l - \tfrac{5}{2})(w + \tfrac{4}{3}).$$

After computing the products on the right and simplifying each equation, we are led to the linear system

$$-\tfrac{2}{3}l + \tfrac{5}{2}w = \tfrac{10}{6} \quad \text{and} \quad \tfrac{4}{3}l - \tfrac{5}{2}w = \tfrac{20}{6}$$

whose unique solution is $l = 15/2$, $w = 8/3$, so $lw = 20$.

20. (A) The side s_k of each square is $1/\sqrt{2}$ times that of the preceding square: $s_k = (1/\sqrt{2})s_{k-1}$, $s_1 = m$ (see figure). The radius r_k of each circle is $1/2$ times the side of its circumscribed square:

$$r_1 = \frac{1}{2}s_1 = \frac{1}{2}m, \ r_2 = \frac{1}{2}s_2 = \frac{1}{2}\frac{1}{\sqrt{2}}m, \ r_k = \frac{1}{2}s_k = \frac{m}{2}\left(\frac{1}{\sqrt{2}}\right)^{k-1}.$$

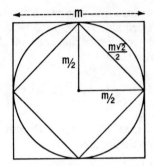

The area of the kth circle is $A_k = \pi(\tfrac{1}{2}m)^2(\tfrac{1}{2})^{k-1}$. The required sum is

$$S_n = A_1 + A_2 + \cdots + A_n$$

$$= \left(\frac{m}{2}\right)^2 \pi + \frac{1}{2}\left(\frac{m}{2}\right)^2 \pi + \frac{1}{4}\left(\frac{m}{2}\right)^2 \pi + \cdots + \frac{1}{2^{n-1}}\left(\frac{m}{2}\right)^2 \pi$$

$$= \frac{m^2\pi}{4}\left[1 + \frac{1}{2} + \cdots + \frac{1}{2^{n-1}}\right] = \frac{m^2\pi}{4}\cdot 2\left[1 - \left(\frac{1}{2}\right)^n\right].$$

As n grows beyond all bounds, $(\tfrac{1}{2})^n$ approaches 0, so that the required sum approaches $m^2\pi/2$.

Part 2

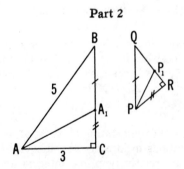

21. (B) Leg $BC = 4$ of the 3, 4, 5 right triangle ABC is divided by the bisector of angle A at A_1 into segments A_1B and A_1C proportional to the adjacent sides AB and AC (see figure):

$$\frac{A_1B}{A_1C} = \frac{A_1B}{4 - A_1B} = \frac{5}{3};$$

so

$$A_1B = \tfrac{5}{2} = PQ \quad \text{and} \quad A_1C = \tfrac{3}{2} = PR$$

are the hypotenuse and leg, respectively, of the second right $\triangle PQR$. Its third side is $RQ = \frac{4}{2} = 2$. Each side of $\triangle PQR$ is one-half the corresponding side of the first right triangle ABC. Also bisector

$$PP_1 = \frac{1}{2} AA_1 = \frac{1}{2} \sqrt{AC^2 + CA_1^2} = \frac{1}{2} \sqrt{3^2 + \left(\frac{3}{2}\right)^2} = \frac{3\sqrt{5}}{4}.$$

22. (A) We are given that $P = QD + R$ and that $Q = Q'D' + R'$. Therefore, by substitution,

$$P = (Q'D' + R')D + R = Q'(DD') + (R + R'D).$$

To see that $R + R'D$ is the remainder in this division by DD', we must show that $R + R'D$ is less than the divisor DD'; we know that $R \leq D - 1$, and $R' \leq D' - 1$ so that

$$R + R'D \leq (D - 1) + (D' - 1)D = DD' - 1 < DD'.$$

23. (B) For real $x > 1$, $6x - 5$ and $2x + 1$ are positive, so that both logarithms are defined. Now

$$\log_3 (6x - 5) - \log_3 (2x + 1)$$
$$= \log_3 \frac{6x - 5}{2x + 1} = \log_3 \frac{6x + 3 - 8}{2x + 1} = \log_3 \left(3 - \frac{8}{2x + 1}\right)$$

approaches $\log_3 3 = 1$ as x increases beyond all bounds, because $8/(2x + 1)$ then approaches 0.

24. (A) The given equation is equivalent to $y = 3(167 - x)/5$. For y to be a positive integer, $(167 - x)$ must be a positive multiple of 5; this is the case for the 33 positive integers $x = 5k + 2$, $k = 0, 1, 2, \cdots, 32$.

OR

We use the theorem stating that $(x, y) = (x_0 - bt, y_0 + at)$ gives all solutions in integers of the equation $ax + by = c$ if a and b are relatively prime integers and (x_0, y_0) is any particular solution, the different integers t giving the different solutions.[†] The theorem applies to the present equation $3x + 5y = 501$ with a particular solution $(x_0, y_0) = (167, 0)$ and all solutions are given by $(x, y) = (167 - 5t, 0 + 3t)$. The integers $t = 1, 2, 3, \cdots, 33$ give all 33 solutions in which both x and y are positive, and only those solutions.

† For a proof of this theorem, see *Continued Fractions* by C. D. Olds, Vol. 9 in this NML series, pp. 44–45.

25. (A) Since p is odd and $p > 1$, $p - 1$ is a positive even integer, say $p - 1 = 2n$; hence $\frac{1}{2}(p - 1) = n$ is a positive integer, and

$$(p - 1)^{(p-1)/2} - 1$$
$$= (2n)^n - 1$$
$$= [(2n) - 1][(2n)^{n-1} + (2n)^{n-2} + \cdots + 2n + 1]$$

always has the factor

$$2n - 1 = (p - 1) - 1 = p - 2$$

as stated in choice (A).

It is easy to see that none of the other alternatives is valid; the odd integer $p = 5$, for instance, constitutes a counterexample to choices (B), (C), (D) and (E).

26. (C) From the given information, we have

$$10^3 = 1000 < 1024 = 2^{10} \quad \text{and} \quad 2^{13} = 8192 < 10,000 = 10^4.$$

Taking common logarithms, we get

$$3 < 10 \log 2, \quad \text{or} \quad \log 2 > \tfrac{3}{10},$$

and

$$13 \log 2 < 4, \quad \text{or} \quad \log 2 < \tfrac{4}{13}.$$

Thus $\log 2$ lies in the interval $(\tfrac{3}{10}, \tfrac{4}{13})$, which is contained in those described by (A) and (B), so (C) is a stronger conclusion than (A) or (B). To eliminate (D) and (E), we note that the inequality $\log 2 < \tfrac{40}{132}$, equivalent to $2^{132} < 10^{40}$, does not follow from the given tabular information.

27. (C) Let the length of a candle be chosen as the unit of length. Let t represent the number of hours before 4 P.M. needed to produce the desired result. In one hour, the faster burning candle shortens by $\frac{1}{3}$, the slower by $\frac{1}{4}$ its length, and in t hours, they shorten by $t/3$ and $t/4$, so their lengths are $1 - t/3$ and $1 - t/4$, respectively. Then $(1 - t/4) = 2(1 - t/3)$, $t = 2\frac{2}{5}$ hours before 4 P.M. The time for the candles to be lighted is therefore $4 - 2\frac{2}{5} = 1\frac{3}{5}$ hours after noon or 1:36 P.M.

28. (E) Denote the set of Mems, Ens and Vees by M, N and V, respectively. Hypotheses I and II tell us only that at least one member of M is not in N, and that N and V are disjoint (that is, have no common members). Suppose sets M and V were identical; then statements (A), (B), (C) would be false.

Now suppose sets M and V were disjoint; then statement (D) would be false. Therefore none of the statements (A), (B), (C), (D) can be deduced from the given hypotheses.

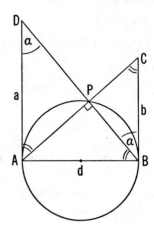

29. (C) Denote the diameter of the circle by d, and the intersection of perpendicular lines AC and BD by P (see figure). Since tangents AD and BC are parallel, $\angle ADB = \angle PBC = \alpha$. The complements of these angles, $\angle ABD$ and $\angle BCA$ are also equal, so right triangles ABD and BCA are similar. The proportionality of corresponding sides then yields

$$\frac{d}{a} = \frac{b}{d} \quad \text{whence} \quad d^2 = ab, \quad d = \sqrt{ab}.$$

OR

After noting that angles D and C are complementary, and that the product of tangents of complementary angles is 1, we may obtain the result from the definitions of the tangents;

$$(\tan C)(\tan D) = \frac{d}{b} \cdot \frac{d}{a} = 1, \quad d^2 = ab, \quad d = \sqrt{ab}.$$

30. (D) The dealer paid d dollars for n radios, so the cost for each radio was d/n dollars. Of these, $n - 2$ were sold for $d/n + 8$ dollars, and 2 were sold for $\frac{1}{2}d/n$, so that the total intake was

$$(n - 2)\left(\frac{d}{n} + 8\right) + 2\frac{d}{2n} = d + 72,$$

that is, 72 dollars more than the cost. This equation reduces to

$$n^2 - 11n = n(n - 11) = d/8,$$

where d and n are positive integers.

For $n \leq 11$, we get a non-positive d.

For $n = 12$, we get $12 \cdot 1 = d/8$, $d = 96$; similarly, every integer greater than 11 yields a positive integer d. Thus 12 is the smallest possible value of n for the given information.

Note. Actually, we need not assume that d is an integer; this follows automatically from the equation

$$d = 8(n^2 - 11n).$$

Part 3

31. (C) Since a and b are consecutive integers, one of them is even, the other odd; hence their product is even. We may let $b = a + 1$. Then $c = ab = a(a + 1) = a^2 + a$ is an even integer and

$$D = a^2 + b^2 + c^2 = a^2 + (a + 1)^2 + a^2(a + 1)^2$$
$$= a^4 + 2a^3 + 3a^2 + 2a + 1 = (a^2 + a + 1)^2$$

is the square of the positive odd integer $a^2 + a + 1 = c + 1$. We conclude that $\sqrt{D} = a^2 + a + 1$ is always an odd positive integer.

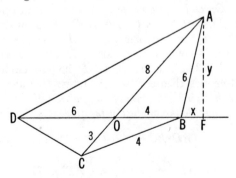

32. (E) Let F denote the foot of the perpendicular from A to diagonal DB extended, and denote BF and FA by x and y respectively (see figure). Then $x^2 + y^2 = 6^2$ and $(x + 4)^2 + y^2 = 8^2$. Subtracting the first of these equations from the second yields

$$8x + 16 = 28, \qquad x = \tfrac{3}{2}.$$

We put $x = \tfrac{3}{2}$ into the first equation and solve for y^2:

$$(\tfrac{3}{2})^2 + y^2 = 6^2, \qquad y^2 = \tfrac{135}{4}.$$

Therefore

$$AD^2 = (10 + x)^2 + y^2 = (\tfrac{23}{2})^2 + \tfrac{135}{4} = \tfrac{664}{4} = 166,$$

$$AD = \sqrt{166}.$$

OR

The law of cosines, applied to triangle AOB yields

$$\cos \sphericalangle AOB = \frac{OA^2 + OB^2 - AB^2}{2(OA)(OB)} = \frac{8^2 + 4^2 - 6^2}{2 \cdot 8 \cdot 4} = \frac{11}{16}.$$

Since angles AOB and AOD are supplementary, $\cos \sphericalangle AOD = -\cos \sphericalangle AOB$, and the Law of Cosines applied to triangle AOD yields

$$AD^2 = OA^2 + OD^2 - 2(OA)(OD) \cos \sphericalangle AOD$$

$$= 8^2 + 6^2 - 2 \cdot 8 \cdot 6(-\tfrac{11}{16}) = 166, \quad AD = \sqrt{166}.$$

OR

The converse of the theorem: *If two chords of a circle intersect, the product of the segments of one is equal to the product of the segments of the other* asserts: *If line segments AC and BD intersect in a point O such that $AO \cdot OC = BO \cdot OD$, then points A, B, C, D lie on a circle*, and is not hard to prove by means of the pairs AOB, DOC and BOC, AOD of similar triangles. In our case, the first pair yields $CD/6 = 3/4$ so that $CD = 9/2$; the second pair yields $AD/BC = 8/4 = 2$ so that $BC = AD/2$. Since A, B, C, D lie on a circle, we may now use Ptolemy's theorem† which states: *If a quadrilateral is inscribed in a circle, the sum of the products of two pairs of opposite sides is equal to the product of the diagonals.* This yields

$$AD \cdot BC + AB \cdot CD = 10 \cdot 11 = 110,$$

$$AD \cdot \frac{AD}{2} + 6 \cdot \frac{9}{2} = \frac{1}{2} AD^2 + 27 = 110,$$

$$AD^2 = 166, \qquad AD = \sqrt{166}.$$

---

† For a proof and a discussion of Ptolemy's theorem and its converse and consequences, see *Geometry Revisited* by H. S. M. Coxeter and S. L. Greitzer, vol. 19 of this NML series, Random House (1967).

33. (D) Let A_1, A_2, A_3 denote the areas of the semicircles with diameters $AB = d_1$, $AC = d_2$, and $CB = d_3$, respectively; let S be the area of the shaded region, and G that of a circle with radius $CD = r$, i.e. $G = \pi r^2$. Then

$$S = A_1 - A_2 - A_3 = \frac{\pi}{8} d_1^2 - \frac{\pi}{8} d_2^2 - \frac{\pi}{8} d_3^2$$

$$= \frac{\pi}{8} [d_1^2 - d_2^2 - d_3^2]$$

$$= \frac{\pi}{8} [(d_2 + d_3)^2 - d_2^2 - d_3^2] \qquad \text{(since } d_2 + d_3 = d_1)$$

$$= \frac{\pi}{4} d_2 d_3.$$

Now $CD = r$ is the altitude of right triangle ADB, hence the mean proportional between d_2 and d_3: $r^2 = d_2 d_3$. It follows that

$$G = \pi r^2 = \pi d_2 d_3 = 4S,$$

and

$$\frac{S}{G} = \frac{1}{4}.$$

Remark: Since the problem does not specify the precise location of the point C on AB, it seems safe to assume that the desired ratio is independent of the position of C. Our calculations are simplified by assuming that C coincides with O, in which case $CD = OA = d_1/2$ and

$$S = \frac{\pi}{2}\left(\frac{d_1}{2}\right)^2 - \frac{\pi}{2}\left(\frac{d_1}{4}\right)^2 - \frac{\pi}{2}\left(\frac{d_1}{4}\right)^2 = \frac{\pi}{16} d_1^2,$$

$$G = \frac{\pi d_1^2}{2} = \frac{\pi d_1^2}{4} = 4S, \quad \text{whence} \quad \frac{S}{G} = \frac{1}{4}.$$

34. (A) Let a, b, c denote the lengths of the sides opposite the vertices A, B, C, and h_a, h_b, h_c those of the altitudes from A, B, C of $\triangle ABC$ (figure on p. 88). Let K, K_O, K_A, K_B, K_C denote the areas of triangles ABC, DEF, ADF, BED, CFE, respectively. The last three have bases $c/(n + 1)$, $a/(n + 1)$, $b/(n + 1)$; their altitudes l_c, l_a, l_b to those bases are parallel to h_c, h_a, h_b respectively, and $l_a/h_a = l_b/h_b = l_c/h_c = n/(n+1)$.

Now

$$K_O = K - K_A - K_B - K_C$$

$$= K - \frac{h_c}{2}\frac{c}{n+1}\frac{n}{n+1} - \frac{h_a}{2}\frac{a}{n+1}\frac{n}{n+1} - \frac{h_b}{2}\frac{b}{n+1}\frac{n}{n+1}$$

$$= K - \frac{n}{(n+1)^2}\left[\frac{1}{2}ch_c + \frac{1}{2}ah_a + \frac{1}{2}bh_b\right]$$

$$= K - \frac{n}{(n+1)^2}3K = K\frac{(n+1)^2 - 3n}{(n+1)^2} = K\frac{n^2 - n + 1}{(n+1)^2}.$$

So

$$\frac{K_O}{K} = \frac{n^2 - n + 1}{(n+1)^2}.$$

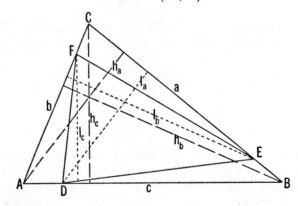

Remark: One could also guess the answer from the obvious fact that, as n approaches zero or infinity, the desired ratio approaches 1. Of the five choices, only (A) has this property.

35. (B) Dividing through by 64 gives the equivalent equation

$$x^3 - \tfrac{9}{4}x^2 + \tfrac{23}{16}x - \tfrac{15}{64} = 0.$$

The negatives of the product and the sum of the roots are the constant term and the coefficient of x^2, respectively. Since the three roots are in arithmetic progression, we may denote them by $(a - d)$, a, $(a + d)$; their sum then is $3a = \tfrac{9}{4}$, so $a = \tfrac{3}{4}$. Their product is $a(a^2 - d^2) = \tfrac{15}{64} = \tfrac{3}{4}(\tfrac{9}{16} - d^2)$ so that $d^2 = \tfrac{1}{4}$, $d = \pm\tfrac{1}{2}$. The difference between the largest and smallest roots is

$$(a + |d|) - (a - |d|) = 2|d| = 2\cdot\tfrac{1}{2} = 1.$$

36. (C) Denote the middle term by a and the ratio by r; then the sum of the five terms, each an integer, is

$$211 = \frac{a}{r^2} + \frac{a}{r} + a + ar + ar^2;$$

r is rational, say $r = c/d$ (where c, d are integers without common divisors), for otherwise ar would not be an integer. Since every term in

$$\frac{ad^2}{c^2} + \frac{ad}{c} + a + \frac{ac}{d} + \frac{ac^2}{d^2} = 211$$

is an integer, c^2 and d^2 both divide a:

$$a = kc^2d^2, \qquad k \text{ an integer.}$$

But then the left side of our equation is divisible by k while 211 is a prime, so $k = 1$ and the equation reduces to

$$d^4 + d^3c + d^2c^2 + dc^3 + c^4 = 211.$$

The integers c and d are both less than 4, since $4^4 = 256 > 211$.
    Neither c nor d can be 1. For, if one of them were, the other would satisfy

$$x^4 + x^3 + x^2 + x + 1 = \frac{x^5 - 1}{x - 1} = 211;$$

but if $x = 2$, the left side is $31 \neq 211$, and if $x = 3$, the left side is $121 \neq 211$.
    Since c and d have no common factor, the only remaining possibility is that one of them is 2, the other 3. Indeed,

$$2^4 + 3 \cdot 2^3 + 3^2 \cdot 2^2 + 3^3 \cdot 2 + 3^4 = 211$$

so that $a = 36$ and r may have the value $\frac{3}{2}$ or $\frac{2}{3}$ (both yield the same terms but in reverse order). The first, third and fifth terms are perfect squares, and their sum is

$$4^2 + 6^2 + 9^2 = 16 + 36 + 81 = 133.$$

37. (A) Let G and M denote the intersection of the medians (the centroid of $\triangle ABC$) and the midpoint of side AC, respectively. Draw MJ perpendicular to line RS at J, and BK and GL parallel to RS intersecting MJ at K and L, respectively. See figure on p. 90. Then

$$MJ = \tfrac{1}{2}(AD + CF) = \tfrac{1}{2}(10 + 24) = 17$$

so that

$$MK = MJ - KJ = MJ - BE = 17 - 6 = 11.$$

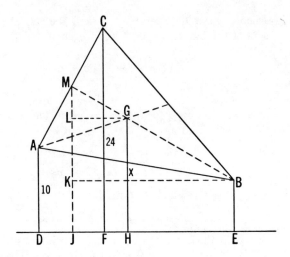

Now since $MG = \frac{1}{3}MB$, $ML = \frac{1}{3}MK$, because the line LG parallel to the base KB of triangle MKB divides the other sides in the same proportion. Therefore the segment sought is

$$x = GH = LJ = MJ - ML = MJ - \frac{1}{3}MK$$
$$= 17 - \frac{1}{3}(11) = \frac{40}{3}.$$

38. (E) Mathematical notation facilitates statements about the system S. Thus let single numbers 1, 2, 3, 4 denote the four distinct pibs (see postulate P_4); these are sets of maas (see P_1). The unique maa common to pibs i and j (see P_2) we denote by ij or ji. Since every maa is one of these (see P_3), the complete set of maas $\{12, 13, 14, 23, 24, 34\}$ contains exactly

$$\binom{4}{2} = \frac{4 \cdot 3}{1 \cdot 2} = \text{six elements } (T_1).$$

The three and only three maas in pib i are the maas ij ($j \neq i$) (T_2). There is exactly one maa in neither pib i nor pib j and hence not in the same pib with ij (T_3). We conclude that all three theorems T_1, T_2, T_3 are deducible from the four postulates P_1, P_2, P_3, P_4 as we have shown above.

Remark: The six intersection points (maas) of four non-parallel coplanar lines, no three of which are concurrent (see figure), serves as a model (finite geometry) which satisfies postulates P_1, P_2, P_3, P_4, where the four pibs are the four sets of three collinear points. We note that in this geometry, two lines (pibs) always intersect in a unique point (maa) but two points (maas) do not always determine a line (pib) of the system.

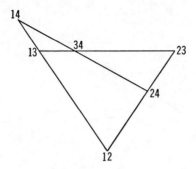

39. (B) The nth set contains n consecutive integers the last of which is the total number of elements in the union of the first n sets; that is, the last integer in S_n is

$$1 + 2 + 3 + \cdots + n = \tfrac{1}{2}n(n+1).$$

The sum S_n may be thought of as starting with this last integer and proceeding downward through n consecutive integers. Thus

$$
\begin{aligned}
S_n = {} & \tfrac{1}{2}n(n+1) \\
& + \tfrac{1}{2}n(n+1) - 1 \\
& + \tfrac{1}{2}n(n+1) - 2 \\
& + \cdots\cdots\cdots \\
& + \tfrac{1}{2}n(n+1) - (n-1) \\
= {} & \tfrac{1}{2}n^2(n+1) - (1 + 2 + \cdots + n - 1) \\
= {} & \tfrac{1}{2}n^2(n+1) - \tfrac{1}{2}n(n-1) \\
= {} & \tfrac{1}{2}n(n^2+1).
\end{aligned}
$$

When $n = 21$ we obtain $S_{21} = \tfrac{1}{2}(21)(21^2 + 1) = 4641$.

Remark: S_n is the sum of n terms of an arithmetic progression with first term $t_1 = \tfrac{1}{2}n(n+1)$ and common difference $d = -1$ so that

$$
\begin{aligned}
S_n &= \tfrac{1}{2}n[2t_1 + (n-1)d] \\
&= \tfrac{1}{2}n[n(n+1) + (n-1)(-1)] \\
&= \tfrac{1}{2}n(n^2+1)
\end{aligned}
$$

as before.

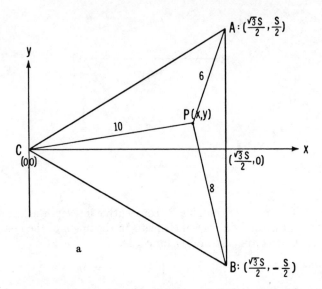

40. (D) Let the vertex C of equilateral $\triangle ABC$ be at the origin of a rectangular coordinate system and the altitude from C coincide with the positive x-axis. Denote the length of one side of $\triangle ABC$ by s; then points A and B have coordinates $(\frac{1}{2}\sqrt{3}s, \frac{1}{2}s)$ and $(\frac{1}{2}\sqrt{3}s, -\frac{1}{2}s)$, respectively (see figure a). Expressions for the squares of the distances from $P(x, y)$ to C, B, and A respectively are

$$x^2 + y^2 = 10^2, \qquad (x - \tfrac{1}{2}\sqrt{3}s)^2 + (y + \tfrac{1}{2}s)^2 = 8^2,$$

and

$$(x - \tfrac{1}{2}\sqrt{3}s)^2 + (y - \tfrac{1}{2}s)^2 = 6^2.$$

Subtracting the third equation from the second gives

$$2sy = 28 \qquad \text{or} \qquad sy = 14.$$

Substituting this value of sy in the second equation and using $x^2 + y^2 = 10^2$ gives

$$10^2 - \sqrt{3}sx + s^2 + 14 = 64, \quad s^2 + 50 = \sqrt{3}sx, \quad sx = \frac{s^2 + 50}{\sqrt{3}}.$$

Substitute the expressions for sx and sy just found into

$$(sx)^2 + (sy)^2 = (x^2 + y^2)s^2$$

and obtain

$$\frac{(s^2 + 50)^2}{3} + 14^2 = 10^2 s^2$$

which reduces to the quadratic equation in s^2

$$s^4 - 200s^2 + 3088 = 0.$$

Its roots are $s^2 = 100 \pm 48\sqrt{3}$, and we discard the smaller because $s^2 > 100$. The desired area is

$$A = \frac{\sqrt{3}s^2}{4} = 25\sqrt{3} + 36 \sim 79 \text{ or choice (D).}$$

OR

We may use the fact that 6, 8, 10 are the sides of a right triangle to facilitate solution of the present problem. To this end, construct $\triangle AP'B$ congruent to $\triangle APC$ (see figure b). Then

$$\sphericalangle PAP' = \sphericalangle PAB + \sphericalangle BAP'$$

$$= \sphericalangle PAB + \sphericalangle CAP = 60°$$

so that isosceles $\triangle PAP'$ is equilateral. Hence $\triangle P'PB$ is a 6, 8, 10 right triangle. Moreover

$$\sphericalangle BPA = \sphericalangle BPP' + \sphericalangle P'PA = 90° + 60° = 150°.$$

Using the law of cosines $\triangle APB$ now yields

$$s^2 = 6^2 + 8^2 - 2 \cdot 6 \cdot 8 \cos 150° = 100 + 48\sqrt{3}.$$

Remark: In our first, algebraic solution, we made no use of the fact that $a = PA$, $b = PB$, $c = PC$ was a Pythagorean triple. The second, geometric solution can also be made to work for any positive triple a, b, c for which the sum of any two is greater than the third. The construction of $\triangle ABP' \sim \triangle ACP$ is as before, $\triangle APP'$ is again equilateral with side a, and $\delta = \sphericalangle P'PB$, now not necessarily a right angle, can nevertheless be determined by the law of cosines since the three sides a, b, c of $\triangle P'PB$ are known. We can now use $\sphericalangle APB = 60° + \delta$ to find s. The calculations are more cumbersome than in the case $a^2 + b^2 = c^2$, but the same principle applies.

In the second solution, $\triangle ABP'$ could have been obtained by a 60° clockwise rotation of AP about A into the position AP'. [See figure b, on p. 94, and cover of this book.]

If, similarly, we rotate BP 60° clockwise about B into BP'' and CP 60° clockwise about C into CP''' we obtain the hexagon $AP'BP''CP'''$, see figure b. The part of the hexagon outside of $\triangle ABC$ consists of the three triangles which have been rotated out of $\triangle ABC$ and which, together, filled up $\triangle ABC$; thus Area of hexagon = 2·Area of $\triangle ABC$.

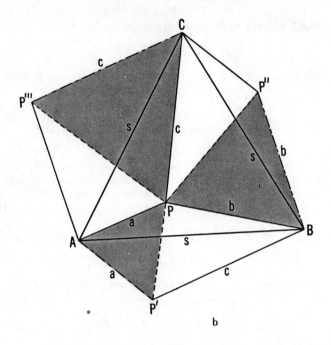

b

On the other hand, the hexagon consists of three equilateral triangles of sides a, b, c respectively (shaded in figure b) and three congruent triangles of sides a, b, and c, whose area can be determined by Heron's formula. Thus

$$2(\text{Area of } \triangle ABC) = \text{Area of hexagon}$$

$$= \frac{\sqrt{3}}{4}(a^2 + b^2 + c^2) + 3\sqrt{\sigma(\sigma - a)(\sigma - b)(\sigma - c)},$$

where $$\sigma = \frac{a + b + c}{2}.$$

In our case, $a = 6$, $b = 8$, $c = 10$; this yields

$$2 \text{ Area}(\triangle ABC) = \frac{\sqrt{3}}{4}(200) + 3\sqrt{12 \cdot 6 \cdot 4 \cdot 2}$$

$$= 50\sqrt{3} + 3 \cdot 24,$$

so Area $(\triangle ABC) = 25\sqrt{3} + 36$.

1968 Solutions

Part 1

1. (D) Let C and d denote the measure of the original circumference and diameter, respectively, so $C = \pi d$. After the increase

$$C + P = \pi(d + \pi) = \pi d + \pi^2 = C + \pi^2.$$

Hence $P = \pi^2$.

2. (B) When the equal numbers $64^{x-1}/4^{x-1} = (64/4)^{x-1} = 16^{x-1}$ and $256^{2x} = (16^2)^{2x} = 16^{4x}$ are expressed as powers of 16 as we have done, the exponents of 16 must be equal, i.e. $4x = x - 1$ so that $x = -\frac{1}{3}$.

OR

Instead of comparing exponents, the logarithms, most conveniently to base 2, 4 or 16, of the equal numbers may be equated. Thus, using 2 as base,

$$\log_2 (64^{x-1}/4^{x-1}) = \log_2 256^{2x}$$

$$(x - 1) \log_2 2^6 - (x - 1) \log_2 2^2 = 2x \log_2 2^8$$

$$(x - 1) \cdot 6 - (x - 1) \cdot 2 = 2x \cdot 8,$$

$x = -\frac{1}{3}$ as before.

3. (A) The slope of the required perpendicular line is -3, the negative reciprocal of the slope $\frac{1}{3}$ of the given line. The line through points $(0, 4)$ and (x, y) has slope $(y - 4)/x$, so the required equation is equivalent to

$$\frac{y - 4}{x} = -3 \quad \text{or} \quad y + 3x - 4 = 0.$$

OR

One may substitute the slope $m = -3$ and the y-intercept $b = 4$ into the form $y = mx + b$ getting the equation $y = -3x + 4$ which is equivalent to $y + 3x - 4 = 0$ of choice (A).

4. (C) Since $a * b = \dfrac{ab}{a+b}$, we have $4 * 4 = \dfrac{4 \cdot 4}{4+4} = \dfrac{16}{8} = 2$, so

$$4 * [4 * 4] = 4 * 2 = \frac{4 \cdot 2}{4+2} = \frac{8}{6} = \frac{4}{3}.$$

5. (A) The definition $f(n) = \frac{1}{3}n(n+1)(n+2)$ gives

$$f(r) = \tfrac{1}{3}r(r+1)(r+2) \quad \text{when} \quad n = r,$$

$$f(r-1) = \tfrac{1}{3}(r-1)r(r+1) \quad \text{when} \quad n = r-1.$$

Subtracting the second from the first, we obtain

$$f(r) - f(r-1) = \tfrac{1}{3}r(r+1)[(r+2) - (r-1)]$$
$$= \tfrac{1}{3}r(r+1)(3) = r(r+1).$$

6. (E) Since the sum of the angles in a triangle is 180° (see diagram on left)

$$\angle E + \angle CDE + \angle DCE = \angle E + S = 180° \quad \text{in} \quad \triangle EDC,$$

and

$$\angle E + \angle BAD + \angle ABC = \angle E + S' = 180° \quad \text{in} \quad \triangle EAB.$$

Hence $S = S' = 180° - \angle E$, so that $r = S/S' = 1$.

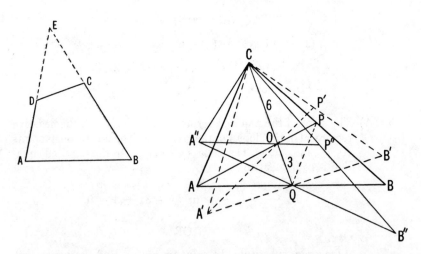

7. (E) To see that the length OP cannot be determined from the given data, we shall show that a segment OP of arbitrary length (and direction) can be constructed. The diagram on right shows several different instances.

Let segment QOC be drawn with $QO = 3$, and hence $OC = 6$ inches. Draw an arbitrary segment OP from O to any point P not on line CQ. Extend PO to the point A such that $OA = 2PO$, and denote by B the point where CP and AQ, extended, intersect. We claim that AP and CQ are medians of $\triangle ABC$. For, triangles AOC and POQ are similar (the lengths of corresponding sides have ratio 2 to 1), so $PQ \parallel AC$, and PQ is half as long as AC. But then P and Q bisect BC and BA, respectively, so that AP and CQ are indeed medians of $\triangle ABC$.

8. (B) Let N represent the positive number, so that the incorrect result is $\frac{1}{6}N$. Since the correct product is $6N$, the error is $6N - \frac{1}{6}N = \frac{35}{6}N$. Therefore the percent error based on the correct result is

$$\frac{\text{Error}}{\text{Correct Result}} \cdot 100 = \frac{3500}{36} \approx 97.$$

9. (E) If $x \geq 2$ or $x \leq -2$, then $x + 2$ and $x - 2$ are both non-negative or both non-positive so that the given equation yields $x + 2 = 2(x - 2)$ or, equivalently, $-(x + 2) = -2(x - 2)$. Hence in this case, $x = 6$. If $-2 < x < 2$, then $x + 2$ is positive and $2(x - 2)$ is negative, so that the given equation yields $x + 2 = -2(x - 2)$, $x = \frac{2}{3}$. The required sum of all values of x satisfying the given equation is $6 + \frac{2}{3} = 6\frac{2}{3}$.

OR

Since the absolute values of two real numbers are equal if and only if their squares are equal, the given equation yields

$$x^2 + 4x + 4 = 4(x^2 - 4x + 4)$$

which, when simplified, yields the quadratic equation

$$3x^2 - 20x + 12 = 0$$

with real roots. The sum of its roots is $-(-20/3) = 6\frac{2}{3}$.

10. (C) First, statements (A) and (B) may be invalid because each requires that the set of all fraternity members be nonempty which is not required by hypothesis I or II. Again, the hypotheses I and II would allow the set of all fraternity members to be a nonempty subset of the set of all students, but neither (D) or (E) permits this and accordingly may be invalid. Now choice (C) is valid under the hypotheses, because by I there exist dishonest students, and by II they cannot be fraternity members.

Part 2

11. (B) Let r_1 and r_2 denote the radii of circles I and II respectively. Equating the arcs, we get

$$\frac{60}{360} \cdot 2\pi r_1 = \frac{45}{360} \cdot 2\pi r_2. \qquad \therefore \ \frac{r_1}{r_2} = \frac{3}{4}.$$

Since the areas of any two circles are proportional to the squares of their radii,

$$\frac{\text{Area (I)}}{\text{Area (II)}} = \frac{r_1^2}{r_2^2} = \frac{3^2}{4^2} = \frac{9}{16}.$$

12. (C) Since $(7\frac{1}{2})^2 + 10^2 = (12\frac{1}{2})^2$ the triangle is a right triangle with the given circle as its circumcircle. The hypotenuse of length $12\frac{1}{2}$ is a diameter of the circle because the right angle opposite it subtends a semicircle. Hence the required radius is one half of that diameter: $\frac{1}{2}(12\frac{1}{2}) = 25/4$.

   Remark: A formula which gives the radius R of the circumcircle directly in terms of the sides a, b, and c of any triangle is

$$R = abc/(4K)$$

   where

$$K = \sqrt{s(s-a)(s-b)(s-c)} = \text{Area } (\triangle ABC),$$

   and

$$s = \tfrac{1}{2}(a+b+c) = \text{semiperimeter of } \triangle ABC.$$

   If you don't identify the given lengths $(\frac{1.5}{2}, \frac{2.0}{2}, \frac{2.5}{2}) = \frac{5}{2}(3,4,5)$ as a Pythagorean triple, you would be obliged to use this formula for R. Use it to check the result in this problem.

13. (B) The sum and the product of roots of the given quadratic equation are $-m$ and n, respectively:

$$m + n = -m \quad \text{and} \quad mn = n.$$

   Hence $m = 1$, $n = -2$, and $m + n = -1$.

14. (E) Choice (E) can be deduced algebraically, for example by writing the given equations in the equivalent form $x - 1 = 1/y$, $y - 1 = 1/x$, whence $y(x-1) = x(y-1) = 1$. Therefore $xy - y = xy - x$, so that $x = y$.

   Remark: Choice (E) was determined without actually finding the values of x (and y) which satisfy the given equations. However, we can easily find them and check our previous result.

Substituting for y (from the second given equation) in the first equation yields

$$x = 1 + \frac{1}{1 + 1/x} = 1 + \frac{x}{x + 1},$$

so

$$x - 1 = \frac{x}{x + 1} \quad \text{and} \quad x^2 - x - 1 = 0.$$

The roots of this quadratic equation are

$$x = \frac{1 + \sqrt{5}}{2} \quad \text{and} \quad x = \frac{1 - \sqrt{5}}{2}.$$

Substituting these into the second equation yields the solutions

$$(x, y) = \left(\frac{1 + \sqrt{5}}{2}, \frac{1 + \sqrt{5}}{2} \right)$$

and

$$(x, y) = \left(\frac{1 - \sqrt{5}}{2}, \frac{1 - \sqrt{5}}{2} \right).$$

In either case, $x = y$.

15. (D) The required product P may be written as

$$P = (2k - 1)(2k + 1)(2k + 3),$$

where k is any positive integer. The integer k has exactly one of the following three properties:

(i)   it is divisible by 3 (i.e. $k = 3m$, m an integer)

(ii) it leaves 1 as remainder when divided by 3 (i.e. $k = 3m + 1$)

(iii) it leaves 2 as remainder when divided by 3 (i.e. $k = 3m + 2$).

In case (i) the last factor of P is divisible by 3. In case (ii) the second factor, $2k + 1 = 2(3m + 1) + 1 = 6m + 3$ is divisible by 3. In case (iii), the first factor, $2k - 1 = 2(3m + 2) - 1 = 6m + 3$, is divisible by 3. Thus, in any event, P is divisible by 3. To see that no larger integer divides all such P, take $P_1 = 1 \cdot 3 \cdot 5$, $P_2 = 7 \cdot 9 \cdot 11$, and observe that 3 is the greatest common divisor of P_1 and P_2.

16. (E) We are told that $-3 < 1/x < 2$. If $x > 0$, taking reciprocals in the inequality on the right yields $x > \frac{1}{2}$. If $x < 0$, multiplication by $-x$ in the left inequality yields $3x < -1$, so $x < -\frac{1}{3}$.

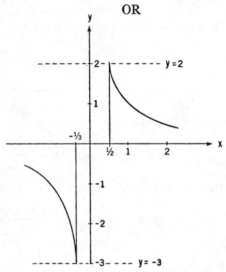

Consider the graph of the function $f(x) = 1/x$ (see figure); it lies in the strip bounded by $y = -3$ and $y = 2$ if and only if $x < -\frac{1}{3}$ or $x > \frac{1}{2}$.

17. (C) The numerator of $f(n)$ is 0 when n is even and -1 when n is odd, because the sum of any two consecutive x's is zero:

$$x_k + x_{k+1} = (-1)^k + (-1)^{k+1} = (-1)^k(1 - 1) = 0,$$

and $(-1)^{-1} = -1$.

Hence $f(2k) = 0$, $f(2k + 1) = -1/(2k + 1)$, and $f(n)$ is therefore contained in $\{0, -1/n\}$.

18. (D) Line AG is a transversal across parallel lines AB and DF, so that the angles FEG and BAE are equal. The former is given equal to angle GEC, which in turn is equal to its vertical angle BEA. Hence $\sphericalangle BAE = \sphericalangle BEA$, so $\triangle ABE$ is isosceles with $BE = BA = 8$. Since corresponding sides of similar triangles DEC and ABC are proportional,

$$\frac{EC}{BC} = \frac{DE}{AB}, \quad \text{that is,} \quad \frac{EC}{8 + EC} = \frac{5}{8}.$$

Thus $8EC = 40 + 5EC$, $3EC = 40$, and $EC = 40/3$.

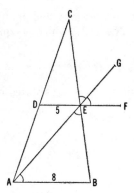

19. (E) Let q and d denote the number of quarters and dimes, respectively, to total \$10. Then, in cents, $25q + 10d = 1000$, which is equivalent to $2d = 5(40 - q)$. Since the left side is a positive even integer, the right side must also be a positive even integer, so $40 - q$ must be even and positive. This is the case when q is any even integer less than 40, so the number n of solutions is 19 or choice (E).

OR

We look for solutions (q, d) in positive integers of the equation $5q + 2d = 200$ (equivalent to the equation above). We observe that $(2, 95)$ is such a solution. By a theorem in number theory† every solution in integers is of the form $(q, d) = (2 + 2t, 95 - 5t)$ where t is any integer. Those integers t for which both q and d are positive are $t = 0, 1, 2, \cdots, 18$ so that the number n of solutions is 19 as found before.

20. (A) The sum of the n interior angles of the polygon in degrees is

$$160 + (160 - 5) + (160 - 5 \cdot 2) + (160 - 5 \cdot 3) +$$

$$\cdots + 160 - 5(n - 1)$$

$$= 160n - 5(1 + 2 + \cdots + n - 1) = 160n - 5\frac{n(n - 1)}{2} \ddagger$$

$$= \frac{5n}{2}[64 - (n - 1)] = \frac{5n}{2}(65 - n).$$

which is equal to $180(n - 2)$ for any convex n sided polygon.

† See *Continued Fractions* by C. D. Olds, Vol. 9 in this NML series, pp. 44–45.

‡ Here we used the fact that the sum of the first k positive integers is $\frac{1}{2}k(k + 1)$, see footnote on p. 114.

Equating these expressions and multiplying by 2/5, we get $n(65 - n) = 72(n - 2)$ which is equivalent to the quadratic equation

$$n^2 + 7n - 144 = 0 \quad \text{or} \quad (n - 9)(n + 16) = 0.$$

Since n is positive, $n = 9$ as stated in choice (A).

Part 3

21. (D) We have

$$
\begin{aligned}
S &= 1! + 2! + 3! + 4! + 5! + \cdots + 99! \\
&= 1 + 2 + 2\cdot3 + 2\cdot3\cdot4 + 2\cdot3\cdot4\cdot5 + \cdots + 99! \\
&= 1 + 2 + 6 + 24 + 10k,
\end{aligned}
$$

k a positive integer, because each of the terms $5!, 6!, \cdots, 99!$ contains the factors 2 and 5 and hence is a multiple of 10. The sum of the units' digits of S is therefore

$$1 + 2 + 6 + 4 = 13.$$

The units' digit in S is accordingly 3 as stated in choice (E).

22. (E) Fundamental in the proof is the fact that a quadrilateral with four given segments as sides exists if and only if the length of each segment is less than the sum of the lengths of the other three.† Now let s_1, s_2, s_3, and s_4 denote the lengths of the four segments. If a quadrilateral exists, then by the fact mentioned above

$$s_1 < s_2 + s_3 + s_4.$$

By hypothesis,

$$s_1 + s_2 + s_3 + s_4 = 1.$$

If we replace the sum $s_2 + s_3 + s_4$ by the smaller number s_1, we obtain the inequality

$$s_1 + s_1 = 2s_1 < 1 \quad \text{so that} \quad s_1 < \tfrac{1}{2}.$$

---

† The "only if" part is obvious since the length of a polygonal path is at least equal to the distance between its endpoints. On the other hand if the length of each segment is less than the sum of the lengths of the other three, label the segments so that $s_1 + s_2 \geq s_3 + s_4$. Then there is a triangle with sides s_1, s_2, $s_3 + s_4$. This triangle may be viewed as a quadrilateral with sides s_1, s_2, s_3, s_4.

Since none of the four segments is in any way special, we can deduce, by the same argument, that

$$s_2 < \tfrac{1}{2}, \quad s_3 < \tfrac{1}{2} \text{ and } s_4 < \tfrac{1}{2}.$$

Conversely, if $s_i < \tfrac{1}{2}$ $(i = 1, 2, 3, 4)$ and $s_1 + s_2 + s_3 + s_4 = 1$, then $s_2 + s_3 + s_4 = 1 - s_1 > 1 - \tfrac{1}{2} = \tfrac{1}{2} > s_1$, so $s_1 < s_2 + s_3 + s_4$. Corresponding inequalities hold for the other segments. Choice (E) is therefore correct. All the other choices fail; for example, a rectangle with adjacent sides of lengths $\tfrac{1}{16}$ and $\tfrac{7}{16}$ has perimeter 1, yet is excluded by all other choices. Choice (D) is clearly excluded, since there is no such division into four segments.

23. (B) The given equality is equivalent to

$$\log (x + 3)(x - 1) = \log (x^2 - 2x - 3);$$

therefore

$$(x + 3)(x - 1) = x^2 - 2x - 3,$$
$$x^2 + 2x - 3 = x^2 - 2x - 3, \quad x = 0.$$

But when $x = 0$, both $x - 1$ and $x^2 - 2x - 3$ are negative so that neither $\log (x - 1)$ nor $\log (x^2 - 2x - 3)$ is defined; thus the equality is satisfied for no real number, as stated in choice (B).

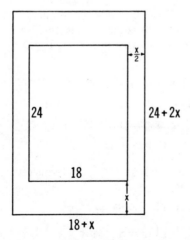

24. (C) Let x and $x/2$ be the width (in inches) of the frame at the top and bottom, and at the sides, respectively. Since the total area is twice the area of the picture, we have

$$(2x + 24)(x + 18) = 2(18)(24)$$

which reduces to

$$2(x^2 + 30x + 216) = 2(2 \cdot 216)$$

or

$$x^2 + 30x - 216 = (x + 36)(x - 6) = 0.$$

Hence $x = 6$ ($x = -36$ is inadmissible). The required ratio is

$$(x + 18)/(2x + 24) = 24/36 = 2/3$$

or choice (C).

25. (C) Let v and vx be the speeds (in yards per unit of time) of Ace and Flash, respectively, and t the time (in the same unit of time) required for Flash to catch Ace. Then the distance in yards run by Flash is $vxt = y + vt$; so that $vt = y/(x - 1)$, and hence the required distance $vtx = xy/(x - 1)$ yards.

Remark: The answer must have the dimension of yards. Since x is dimensionless and y is measured in yards, only choices (A) and (C) meet this requirement. We can eliminate (A) by noting that as x approaches 1, the solution must approach infinity.

26. (E) The sum $S = 2 + 4 + \cdots + 2K = 2(1 + 2 + \cdots + K)$ $= K(K + 1).$† When $K = 999$, $S = 999,000 < 1,000,000$; but when $K = 1,000$, $S = 1000 \cdot 1001 = 1,001,000 > 1,000,000$ so that $N = 1,000$ is the smallest integer for which $S > 1,000,000$. The sum of the digits in 1,000 is 1.

27. (B) When n is even, grouping the n terms into $\frac{1}{2}n$ pairs gives

$$S_n = (1 - 2) + (3 - 4) + \cdots + [(n - 1) - n]$$
$$= \underbrace{-1 - 1 \cdots \cdots - 1}_{n/2 \text{ terms}} = -\tfrac{1}{2}n.$$

When n is odd, grouping the terms after the first into $\frac{1}{2}(n - 1)$ pairs gives

$$S_n = 1 + (-2 + 3) + (-4 + 5) + \cdots + [-(n - 1) + n]$$
$$= 1 + \underbrace{[1 + 1 + \cdots + 1]}_{(n - 1)/2 \text{ terms}} = 1 + \tfrac{1}{2}(n - 1) = \tfrac{1}{2}(n + 1).$$

Hence $S_{17} + S_{33} + S_{50} = \frac{18}{2} + \frac{34}{2} - \frac{50}{2} = 9 + 17 - 25 = 1.$

† The sum of the first K positive integers is $\frac{1}{2}K(K + 1)$, see footnote on p. 114.

28. (D) We are given that $\frac{1}{2}(a + b) = 2\sqrt{ab}$ which is equivalent, after dividing by b and multiplying by 2, to $a/b + 1 = 4\sqrt{a/b}$; this, in turn, is equivalent to the following quadratic equation in $\sqrt{a/b}$:

$$\left(\sqrt{\frac{a}{b}}\right)^2 - 4\sqrt{\frac{a}{b}} + 1 = 0, \quad \text{so} \quad \sqrt{\frac{a}{b}} = \frac{4 \pm \sqrt{12}}{2} = 2 \pm \sqrt{3}.$$

The solution $2 - \sqrt{3}$ must be rejected because the requirement $a > b > 0$ implies $a/b > 1$ while $2 - \sqrt{3} < 1$. Hence

$$\frac{a}{b} = (2 + \sqrt{3})^2 = 7 + 4\sqrt{3} \approx 7 + 6.928 \approx 14.$$

29. (A) We shall use the fact that for $0 < x < 1$, any positive power of x is less than one.† In particular, $y = x^x < 1$. Moreover,

$$\frac{x}{y} = \frac{x}{x^x} = x^{1-x} < 1$$

since $1 - x > 0$, so $x < y$; and

$$\frac{z}{y} = \frac{x^y}{x^x} = x^{y-x} < 1,$$

since $y - x > 0$; hence $z < y$. Finally

$$\frac{x}{z} = \frac{x}{x^y} = x^{1-y} < 1, \quad \text{since} \quad 1 - y > 0.$$

$\therefore x < z$. It follows that $x < z < y$.

OR

Since $0 < x < 1$, $\log x < 0$. If an inequality is multiplied by a negative number, the sign of the inequality is reversed; thus, multiplying $0 < x < 1$ by $\log x$ yields

$$0 > x \log x > \log x, \quad \text{or} \quad 0 > \log x^x = \log y > \log x.$$

Since the logarithmic function is an increasing function, and since $0 = \log 1$, it follows that $1 > y > x$. Again, multiplying by $\log x$, we obtain

$$\log x < y \log x < x \log x, \quad \text{or} \quad \log x < \log x^y = \log z < \log y,$$

so $x < z < y$.

---

† Since $x < 1$, we have $\log x < 0$. Hence if $t > 0$, then $\log x^t = t \log x < 0$, so that $x^t < 1$.

30. (A) A convex point set, by definition, is such that with every pair of
its points it contains the entire line segment joining them. Conse-
quently, a polygon is convex (i.e. bounds a convex set) if and
only if all its angles are $\leq 180°$. It follows from the convexity
of P_2 that each side of P_1 can intersect P_2 in at most two
points, so that the total number of intersections is at most $2n_1$.
We show next that this maximum of $2n_1$ intersections is
always attainable by a suitably determined P_2 which passes
through any two preassigned interior points on each of the n_1
sides of P_1. First mark these $2n_1$ points. Next join the two
points adjacent to each vertex of P_1 by a "cutoff" segment.
Let the two "cutoff" segments through the two marked points
on any side of P_1 be extended outside P_1 from the two points.
The intersection of these two "cutoff" extensions is taken to be
a vertex of P_2 if they meet, but if not, a vertex of P_2 may
be assigned as any point on each of the extensions. It is convenient in the latter case to take the two vertices of P_2 as
consecutive and joined by a side of P_2 parallel to the side of P_1
being considered. Each "cutoff" segment with its two-way
extensions to consecutive vertices of P_2 is a side of P_2 with two
points of P_1 on its interior. There are n_1 "cutoff" segments
(one for each vertex of P_1) and hence P_2 intersects P_1 in
exactly the maximum of $2n_1$ points. Polygon P_2 is convex
because each of its angles is either an angle of a triangle or
equal to an exterior angle of a triangle, hence in any case less
than $180°$.

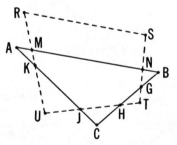

Remark 1: In the figure, P_1 is $\triangle ABC$ and P_2 is a convex quadrilateral $RSTU$ intersecting P_1 in the maximum number
$2n_1 = 6$ points G, H, J, K, M, N which lie on "cutoff"
extensions of segments NG, HJ, KM. The vertices of P_2
on these extensions are $S, T; T, U; U, R$, respectively.
Side RS of P_2 is taken parallel to side AB of P_1
because the "cutoff" extensions outside of P_1 at M and N do
not meet, so that R and S are taken as consecutive vertices of
P_2 on extensions MR and NS. Angles R and S are equal to
exterior angles of "cutoff" triangles AKM and BGN, respec-

tively, while angles T and U are angles in triangles GHT and JKU, respectively.

Remark 2: When $n_1 = 1$ and $n_2 = 3$, P_1 is a segment while P_2 is a triangle. The maximum number of intersections is 2, eliminating choices (B), (C), and (D).

Part 4

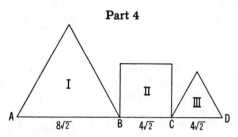

31. (D) Using the area formula $K = (\sqrt{3}/4)s^2$ for an equilateral triangle in terms of a side s in triangles I and III, we obtain the lengths $AB = 8\sqrt{2}$ and $CD = 4\sqrt{2}$. The length of a side of the square II is $BC = 4\sqrt{2}$ so that the length of AD is $16\sqrt{2}$; and $12\frac{1}{2}\%$ or $\frac{1}{8}$ of this is $2\sqrt{2}$ and is to be subtracted from BC, reducing the length of BC to $2\sqrt{2}$ or $\frac{1}{2}$ of its former value. Hence the area of the square is reduced to $\frac{1}{4}$ of its former value. This is a reduction of 75% in the area of the square.

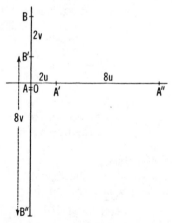

32. (C) Let (u, v) denote the uniform speeds of (A, B) in yards per minute. Then their equal distances from O after 2 and 10 minutes (represented by $OA' = OB'$ and $OA'' = OB''$ in the figure) may be expressed in terms of u and v as

$$2u = 500 - 2v \quad \text{and} \quad 10u = 10v - 500,$$

respectively. Adding these two equations gives $12u = 8v$, so that $u/v = 2/3$, and the ratio of the speeds of A and B is $u{:}v = 2{:}3$ or choice (C).

Remark: The positions of A and B after two minutes are A' and B' with $OA' = 2u = OB'$. During the next 8 minutes, A moves $8u$ more yards to A'', B moves $8v$ more yards to B'', and

$$OA'' = 10u = OB'' = 8v - 2u,$$

whence

$$12u = 8v \quad \text{and} \quad u/v = 2/3.$$

Note that this solution makes no use of the information that B, initially, is 500 yards away from O; but we did use the fact that B was walking *towards* O (otherwise their equal distance from O at two different times might have led us to believe that $u = v$). To find the ratio u/v, one relation involving u and v suffices, and the solution confirms this principle.

33. (A) Let x, y, and z denote the first, second, and third digits of N in base 9 so that

$$81x + 9y + z = 49z + 7y + x \quad \text{or} \quad y = 8(3z - 5x).$$

Since $0 \leq y < 7$ (it appears as a digit in base 7), the integer $n = 3z - 5x$ is zero (otherwise $|8n|$ would be greater than 7). Hence y, the middle digit, is zero. Moreover $0 < z < 7$ (since N has three digits in base 7); and since $3z = 5x$, z is divisible by 5. Hence $z = 5$, and $x = 3$, so that

$$N = 305_9 = 503_7 = 248_{10}.$$

34. (B) Let d and p denote the number of votes which originally defeated and later passed the bill, respectively. Then $400 - d$ and $400 - p$ were the number of votes first for, later against the bill, respectively, and $d - (400 - d) = 2d - 400$ was the margin of defeat in the first vote, while $2p - 400$ was the margin of passage in the second. The two pieces of information given in the problem lead to the equations

$$2p - 400 = 2[2d - 400] \quad \text{and} \quad p = \tfrac{12}{11}d.$$

The first is equivalent to $2d - p = 200$, and after substituting for p from the second, we find $d = 220$, $p = 240$. The required difference is $p - (400 - d) = 240 - 180 = 60$.

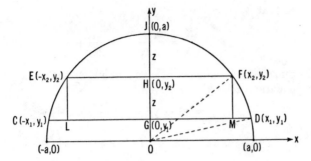

35. **(D)** Using the area formulas for a trapezoid and a rectangle, we write

$$\frac{K}{R} = \frac{(1/2)HG(EF+CD)}{HG \cdot EF} = \frac{1}{2}\left(1 + \frac{CD}{EF}\right) = \frac{1}{2} + \frac{1}{2}\frac{CD}{EF};$$

see the figure. Now

$$\frac{CD}{EF} = \frac{2GD}{2HF} = \frac{GD}{HF} = \frac{\sqrt{OD^2 - OG^2}}{\sqrt{OF^2 - OH^2}} = \frac{\sqrt{a^2 - OG^2}}{\sqrt{a^2 - OH^2}}$$

by use of the Pythagorean Theorem applied to right triangles OGD and OHF. If we denote JH by z so that $JG = 2z$, then $OH = a - z$, $OG = a - 2z$, and

$$\frac{CD}{EF} = \frac{\sqrt{a^2 - (a-2z)^2}}{\sqrt{a^2 - (a-z)^2}} = \frac{\sqrt{4az - 4z^2}}{\sqrt{2az - z^2}} = \frac{\sqrt{4a - 4z}}{\sqrt{2a - z}}.$$

Now as OG approaches a, z approaches O, so that CD/EF approaches $\sqrt{4a/(2a)} = \sqrt{2}$.

Hence $K/R = \frac{1}{2} + \frac{1}{2}(CD/EF)$ approaches $\frac{1}{2} + \frac{1}{2}\sqrt{2} = \frac{1}{2} + 1/\sqrt{2}$, as stated in choice (D).

OR

The analytic version of the solution is this: Choose O as the origin and OJ as the y-axis of a Cartesian coordinate system. Denote the coordinates of points D, C, F, E, G as shown in the figure; since H is the midpoint of GJ, its ordinate y_2 is the average $(y_1 + a)/2$ of the ordinates of G and J. Now the area formulas yield

$$K = \frac{2x_1 + 2x_2}{2}\left(\frac{y_1 + a}{2} - y_1\right) = (x_1 + x_2)\left(\frac{a - y_1}{2}\right)$$

$$R = 2x_2\left(\frac{y_1 + a}{2} - y_1\right) = x_2(a - y_1),$$

so that

$$\frac{K}{R} = \frac{x_1 + x_2}{2x_2} = \frac{x_1}{2x_2} + \frac{1}{2}.$$

Since the coordinates of every point on the circle satisfy the equation $x^2 + y^2 = a^2$, we have

$$x_1^2 = a^2 - y_1^2 = (a - y_1)(a + y_1),$$

$$(2x_2)^2 = 4x_2^2 = 4a^2 - 4y_2^2 = 4a^2 - 4\left(\frac{y_1 + a}{2}\right)^2$$

$$= 4a^2 - [y_1^2 + 2ay_1 + a^2] = 3a^2 - 2ay_1 - y_1^2$$

$$= (a - y_1)(3a + y_1).$$

Thus

$$\left(\frac{x_1}{2x_2}\right)^2 = \frac{(a - y_1)(a + y_1)}{(a - y_1)(3a + y_1)} = \frac{a + y_1}{3a + y_1};$$

as $y_1 \to a$, this fraction approaches

$$\frac{a + a}{3a + a} = \frac{2}{4} = \frac{1}{2}$$

and $x_1/2x_2$ approaches $1/\sqrt{2}$. Hence

$$\frac{K}{R} \to \frac{1}{\sqrt{2}} + \frac{1}{2}.$$

1969 Solutions

Part 1

1. (B) We are asked to solve the equation

$$\frac{a + x}{b + x} = \frac{c}{d}.$$

An equivalent equation is

$$ad + xd = bc + xc, \quad \text{so that} \quad (c - d)x = ad - bc,$$

and hence

$$x = \frac{ad - bc}{c - d}.$$

Comment: When c/d is chosen as (i) -1, (ii) the reciprocal of a/b, (iii) the square of a/b, then $-x$ takes the values

(i) $\frac{1}{2}(a + b)$, (ii) $a + b$, (iii) $ab/(a + b)$, respectively;

i.e. (i) the arithmetic mean, (ii) the sum, (iii) half the harmonic mean of the numbers a and b.

2. (A) Let C represent the cost in dollars. Then

$$x = C - .15C = .85C, \quad \text{and} \quad y = C + .15C = 1.15C.$$

Therefore the required ratio is

$$y/x = 1.15C/.85C = 23/17.$$

3. (E) The following identity is valid whenever n and r are integers with $n > r$; it can be verified by direct multiplication:

$$x^n - x^r = (x - 1)(x^{n-1} + x^{n-2} + \cdots + x^r).$$

When we set $x = 2$, we get $x - 1 = 1$ so that

$$2^n - 2^r = 2^{n-1} + 2^{n-2} + \cdots + 2^r.$$

Now $N = 11000_2 = 2^4 + 2^3$, and, by means of above identity with $n = 3$, $r = 0$, we get

$$N - 1 = 2^4 + (2^3 - 1) = 2^4 + (2^2 + 2 + 1) = 10111_2$$

as stated in choice (E).

OR

We can just subtract, using the familiar algorithm, modified for base 2:

$$
\begin{array}{r}
11000 \\
-1 \\
\hline
10111
\end{array}.
$$

OR

We can convert N to base 10 (getting $N = 24$), then subtract 1, then convert 23 back to base 2, getting $N - 1 = 10111_2$.

4. (E) By definition, $(3, 2) * (0, 0) = (3 - 0, 2 + 0) = (3, 2)$, and $(x, y) * (3, 2) = (x - 3, y + 2)$. If $(3, 2)$ and $(x - 3, y + 2)$ represent identical pairs, then $3 = x - 3$, so that $x = 6$ as stated in choice (E). Incidentally, $2 = y + 2$ yields $y = 0$.

5. (B) The sum of all possible values of N such that

$$
N - \frac{4}{N} = R
$$

is the sum of the distinct roots of the equivalent quadratic equation

$$
N^2 - RN - 4 = 0.
$$

These roots are

$$
N = \frac{R + \sqrt{R^2 + 16}}{2} \quad \text{and} \quad N = \frac{R - \sqrt{R^2 + 16}}{2};
$$

their sum is R.

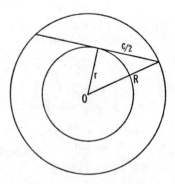

6. (C) From the common center O of the circles, draw the radius r of the smaller circle to its point of tangency with the chord c,

and draw a radius R of the larger circle to an endpoint of the chord. For the right triangle with legs r, $c/2$ and hypotenuse R, the Pythagorean theorem gives

$$\left(\frac{c}{2}\right)^2 = R^2 - r^2,$$

while the area of the ring is

$$\tfrac{25}{2}\pi = \pi R^2 - \pi r^2 = \pi(R^2 - r^2).$$

Therefore $\left(\dfrac{c}{2}\right)^2 = \dfrac{25}{2}$, $c^2 = 50$, and $c = 5\sqrt{2}$.

7. (A) Since the points $(1, y_1)$ and $(-1, y_2)$ lie on the graph of $y = ax^2 + bx + c$, we can substitute their coordinates into the equation, getting

$$y_1 = a + b + c \qquad \text{and} \qquad y_2 = a - b + c.$$

Therefore $y_1 - y_2 = 2b = -6$, and $b = -3$.

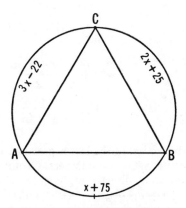

8. (D) Arcs AB, BC, and CA subtend the inscribed interior angles C, A, and B of $\triangle ABC$ which therefore have measures $\frac{1}{2}(x + 75°)$, $\frac{1}{2}(2x + 25°)$, and $\frac{1}{2}(3x - 22°)$, respectively, with sum 180°. The resulting equation

$$\tfrac{1}{2}(x + 75°) + \tfrac{1}{2}(2x + 25°) + \tfrac{1}{2}(3x - 22°) = 180°$$

has the solution $x = 47°$ from which we find

$$(\measuredangle C, \measuredangle A, \measuredangle B) = (61°, 59\tfrac{1}{2}°, 59\tfrac{1}{2}°),$$

so that $\measuredangle C$ is an interior angle of 61°.

9. (C) The integers form an arithmetic progression with first term $a = 2$, common difference $d = 1$, and number of terms $n = 52$. Hence their sum is

$$S = \tfrac{1}{2}n[2a + (n-1)d] = \tfrac{1}{2}52[4 + 51] = \tfrac{1}{2} \cdot 52(55)$$

and their average $S/52 = \tfrac{1}{2} \cdot 55 = 27\tfrac{1}{2}$ or choice (C).

Comment: Virtually all properties of the sum of an arithmetic progression are based on the identity

(*) $$1 + 2 + 3 + \cdots + n = \tfrac{1}{2}n(n+1).†$$

In particular, *the arithmetic mean of n numbers in arithmetic progression is the arithmetic mean of the smallest and largest.*

Proof: Let the n numbers, in increasing order, be

$$a, \; a + d, \; a + 2d, \; \cdots, \; a + (n-1)d.$$

Their arithmetic mean is

$$\frac{1}{n}[a + a + d + a + 2d + \cdots + a + (n-1)d]$$

$$= \frac{1}{n}[na + d(1 + 2 + \cdots + n - 1)]$$

$$= a + \frac{d}{n}\frac{(n-1)n}{2} \qquad \text{(by identity (*))}$$

$$= \frac{1}{2}[a + a + (n-1)d],$$

where we recognize, in the last expression, the arithmetic mean of the first and last of our numbers. (Note: an alternative proof can be obtained by directly applying the Gauss trick to the progression $a, \; a + d, \; \cdots, \; a + (n-1)d$.)

Applying this to the problem at hand, we get

$$\tfrac{1}{2}[2 + 53] = 27\tfrac{1}{2}.$$

---

† This formula for the sum S_n of the integers from 1 to n is derived by the famous Gauss trick of writing the sum twice, the second time in reverse order, and adding:

$$S_n = 1 \qquad + 2 \qquad + 3 + \cdots \qquad + n-1 \; + n$$
$$S_n = n \qquad + n - 1 + n - 2 + \cdots + 2 \qquad + 1$$

$$2S_n = n + 1 + n + 1 + n + 1 + \cdots + n + 1 + n + 1;$$

since there are n terms on the right, all equal to $n + 1$, we obtain $2S_n = n(n + 1)$.

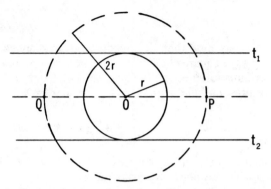

10. (C) Let O denote the center, r the radius, and t_1 and t_2 the two parallel tangents to the given circle (see figure). Then the locus of all points equidistant from t_1 and t_2 is the parallel line QOP midway between them (and hence through O). The intersections P and Q of this locus with the concentric circle of radius $2r$ (dotted in figure) together with the center O itself are the only three points which are equidistant from the circle and its two parallel tangents t_1 and t_2. The number of such points is 3.

Part 2

11. (B) Since the sum of segments $PR + RQ$ is to be a minimum, points P, R, and Q must be collinear, so that the quotient of the difference of the y's and the difference of the x's, in the same order, is constant for any pair of the points. Using the pairs P and R, and P and Q gives

$$\frac{m - (-2)}{1 - (-1)} = \frac{2 - (-2)}{4 - (-1)} \quad \text{which is equivalent to } m = -\tfrac{2}{5}.$$

Comment: In a rectangular coordinate system, the quotients equated above are called the slope of the line.

12. (A) The expression F may be written as

$$F = x^2 + \frac{8}{3}x + \frac{m}{2} = \left(x^2 + \frac{8}{3}x + \frac{16}{9}\right) + \left(\frac{m}{2} - \frac{16}{9}\right)$$

$$= \left(x + \frac{4}{3}\right)^2 + \left(\frac{m}{2} - \frac{16}{9}\right),$$

which is the square of the linear expression $(x + \tfrac{4}{3})$ in x provided $m/2 - 16/9 = 0$ or $m = 32/9$, a particular value of m between 3 and 4 as required in choice (A).

OR

We may note that any quadratic expression $ax^2 + bx + c$ in x is the square of a linear expression in x if and only if its discriminant $b^2 - 4ac$ is equal to zero. Here F is a quadratic expression in x with $a = 1$, $b = \frac{8}{3}$, $c = m/2$, so that

$$b^2 - 4ac = \frac{64}{9} - 4\left(\frac{m}{2}\right) = 0 \text{ or } m = \frac{32}{9}.$$

Hence $m = \frac{32}{9}$ is the only value of m for which F is the square of a linear expression in x.

13. (B) The difference of the areas of the larger and smaller circles, $\pi R^2 - \pi r^2$, is equal to the area of the region outside the smaller and inside the larger circle. By hypothesis, this area times a/b is equal to the area of the larger circle:

$$\pi R^2 = \frac{a}{b}(\pi R^2 - \pi r^2) \quad \text{or} \quad \frac{a}{b}r^2 = R^2\left(\frac{a}{b} - 1\right).$$

Hence

$$\frac{R^2}{r^2} = \frac{a}{a - b}$$

and the required ratio is

$$\frac{R}{r} = \frac{\sqrt{a}}{\sqrt{a - b}}.$$

14. (A) Since the fraction $(x^2 - 4)/(x^2 - 1)$ is required to be positive, $x^2 - 4$ and $x^2 - 1$ must be both positive or both negative. When $x^2 - 4 > 0$ so that $|x| > 2$, then $x^2 - 1 = (x^2 - 4) + 3 > 0$ also. When $x^2 - 1 < 0$ so that $|x| < 1$, then $x^2 - 4 = (x^2 - 1) - 3 < 0$ also. The set of all values of x for which the given fraction is positive is $x > 2$ or $x < -2$ or $-1 < x < 1$.

Remark: The fraction $f(x) = (x^2 - 4)/(x^2 - 1)$ is an *even function* of x; this means $f(x)$ has the property

$$f(-x) = f(x)$$

with the geometric consequence that the graph of $f(x)$ is symmetric with respect to the y-axis. In particular, $f(x) > 0$ on symmetrically located points of the domain of f, and our solution confirms this fact.

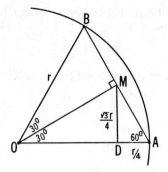

15. (D) Since chord AB has length r, $\triangle AOB$ is equilateral. Perpendicular OM bisects AB, so that AM has length $r/2$ and is the hypotenuse of right triangle MDA with legs $AD = \frac{1}{2}AM = r/4$, and $DM = \sqrt{3}r/4$ (because $\angle A = 60°$).

$$\therefore \text{ Area of } \triangle MDA = \tfrac{1}{2}(AD)(DM) = \frac{1}{2}\cdot\frac{r}{4}\cdot\frac{\sqrt{3}r}{4} = \frac{r^2\sqrt{3}}{32}.$$

OR

Triangles MAD and OAM are similar (both having angles of $30°$, $60°$ and $90°$); and AD and AM are corresponding sides. Since

$$\frac{AD}{AM} = \frac{1}{2}, \qquad \frac{\text{Area } \triangle MAD}{\text{Area } \triangle OAM} = \left(\frac{1}{2}\right)^2 = \frac{1}{4}.$$

$$\text{Area } \triangle OAM = \frac{1}{2}AM\cdot OM = \frac{1}{2}\frac{r}{2}\frac{r\sqrt{3}}{2} = \frac{r^2\sqrt{3}}{8}.$$

$$\therefore \text{ Area } \triangle MAD = \frac{1}{4}\frac{r^2\sqrt{3}}{8} = \frac{r^2\sqrt{3}}{32}.$$

16. (E) In the binomial expansion

$$(a - b)^n = a^n - na^{n-1}b + \frac{n(n-1)}{2}a^{n-2}b^2 - \cdots,$$

the sum of the second and third terms, when $a = kb$, is

$$-n(kb)^{n-1}b + \frac{n(n-1)}{2}(kb)^{n-2}b^2 = 0,$$

which yields, after division by $nk^{n-2}b^n$,

$$-k + \frac{n-1}{2} = 0 \quad \text{or} \quad n = 2k+1.$$

17. (D) The left member of the given equation can be factored to give the equivalent equation

$$(2^x - 2)(2^x - 6) = 0.$$

Since a product is zero if and only if at least one of the factors is zero,

$$2^x - 2 = 0, \quad 2^x = 2, \quad x = 1,$$

or

$$2^x - 6 = 0, \quad 2^x = 6, \quad 2^{x-1} = 3, \quad (x - 1)\log 2 = \log 3,$$

$$x - 1 = \frac{\log 3}{\log 2}, \qquad x = 1 + \frac{\log 3}{\log 2}$$

which is the value of x stated in choice (D). The value $x = 1$ satisfies the given equation also.

18. (B) Each of the graphs consists of a pair of nonparallel straight lines, the first pair having equations

$$\text{I: } x - y + 2 = 0 \qquad \text{and} \qquad \text{II: } 3x + y - 4 = 0$$

and the second pair having equations

$$\text{III: } x + y - 2 = 0 \qquad \text{and} \qquad \text{IV: } 2x - 5y + 7 = 0.$$

The intersections of line III with I and II are the two distinct points $(0, 2)$ and $(1, 1)$, and those of line IV with I and II are two more distinct points $(-1, 1)$ and $(\frac{13}{17}, \frac{29}{17})$, giving a total of four distinct points common to the graphs of the two given equations.

Comment: There can be no more than four points common to the two graphs, because each line of the first (second) pair can intersect those of the second (first) pair in no more than two distinct points. It is interesting to note that in the present problem, the slight change of equation IV to $x - 5y + 4 = 0$ results in the two graphs having only three distinct points in common, even though each line of the second pair still intersects the first pair in two distinct points.

Question: The graphs of the four lines involved in the original problem intersect in six points. Can you explain the two extra points?

19. (B) The given equation $x^4y^4 - 10x^2y^2 + 9 = 0$ is equivalent to $(x^2y^2 - 1)(x^2y^2 - 9) = 0$. The product on the left side is zero only if $x^2y^2 - 1 = 0$, i.e. $x^2y^2 = 1$, or if $x^2y^2 - 9 = 0$, i.e.

$x^2y^2 = 9$. Therefore $xy = \pm 1$ or $xy = \pm 3$. The ordered pairs having positive integral values are $(x, y) = (1, 1)$, $(1, 3)$, $(3, 1)$. There are 3 such pairs as stated in choice (B).

20. **(C)** Let x denote the first and y the second factor in the proposed product P. Then

$$3.6 \cdot 10^{18} < x < 3.7 \cdot 10^{18}, \quad \text{and} \quad 3.4 \cdot 10^{14} < y < 3.5 \cdot 10^{14}.$$

Member by member multiplication of these inequalities gives

$$(3.6)\,(3.4)\,10^{32} < xy < (3.7)\,(3.5)\,10^{32}.$$

Since the lower bound of $P = xy$ on the left and the upper bound on the right are both 34 digit numbers, P is also a 34 digit number as stated in choice (C).

Part 3

21. **(E)** The distance from the line $x + y = \sqrt{2m}$ to any point (x_1, y_1) is

$$d = \left| \frac{x_1 + y_1}{\sqrt{2}} - \sqrt{m} \right|.$$

Its distance from the origin, $(0, 0)$, is therefore $\sqrt{m}$. Thus it is tangent to the circle $x^2 + y^2 = m$, where m may be any non-negative number.

OR

Using the given linear equation, we find $y = \sqrt{2m} - x$ and substitute it into the given quadratic equation:

$$x^2 + y^2 = x^2 + (\sqrt{2m} - x)^2$$
$$= 2x^2 - 2x\sqrt{2m} + 2m = m,$$

$$x^2 - x\sqrt{2m} + \frac{m}{2} = 0, \quad \left(x - \sqrt{\frac{m}{2}} \right)^2 = 0, \quad x = \sqrt{\frac{m}{2}} = y.$$

Note that $x = \sqrt{m/2}$ is always a double root of this quadratic equation. This shows that the two graphs have a single point in common, hence are tangent no matter what non-negative value m has.

Comment: Let the graphs of $x^2 + y^2 = m$ and $x + y = \sqrt{2m}$ be magnified by a factor k [i.e. each point (x, y) is replaced by (kx, ky)]. The resulting point sets satisfy the equations $x^2 + y^2 = k^2m$ and $x + y = k\sqrt{2m} = \sqrt{2k^2m}$, respectively. This pair differs from the original pair in that the number m has been replaced by k^2m. If the original graphs were tangent, so are the magnified versions. Since k^2 is an arbitrary positive constant, we conclude: If the two given graphs were tangent for some non-negative number m_1, they are tangent for any other non-negative number m_2. To see this, set $m_2/m_1 = k^2$ so that $m_2 = k^2m_1$.

This reasoning enables us to eliminate choices (A), (B), (C), (D) immediately. If, in addition, we assume that one of the choices offered is correct, then (E) must be it.

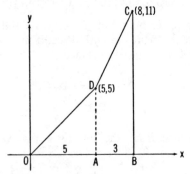

22. (C) The area whose measure K is required consists of the isosceles right triangle with legs OA and AD of length 5 and the trapezoid $ABCD$ with altitude $AB = 3$ and bases $BC = 11$ and $AD = 5$. Thus

$$K = \text{Area of } \triangle OAD + \text{Area of trapezoid } ABCD$$

$$= \tfrac{1}{2}5 \cdot 5 + \tfrac{1}{2}3(5 + 11) = 36.5.$$

23. (A) For $k = 2, 3, \cdots, n - 1$, the sum $n! + k$ is divisible by k because $n!$ has k as a factor. Therefore, any integer m such that $n! + 1 < m < n! + n$ is composite, and there are no prime numbers greater than $n! + 1$ and less than $n! + n$.

24. (E) Let Q, Q' and Q'' be the quotients in the respective divisions of P, P' and RR' by D, so that

$$P = QD + R, \qquad P' = Q'D + R', \qquad RR' = Q''D + r'.$$

Multiplication of the first two equations followed by replacement of RR' from the third gives

$$PP' = (QD + R)(Q'D + R')$$

$$= (QQ'D + QR' + Q'R)D + RR'$$

$$= (QQ'D + QR' + Q'R + Q'')D + r'.$$

Since $r' < D$ and division is unique, the remainder r in the division of PP' by D is equal to r' as stated in choice (E).

25. (D) $\log_2 a + \log_2 b = \log_2 ab \geq 6$. Since the logarithmic function is an increasing function, it follows that

$$ab \geq 2^6.$$

We complete the problem with the help of the arithmetic-geometric mean inequality (proved below); *the geometric mean $\sqrt{ab}$ of two positive numbers a, b does not exceed their arithmetic mean $(a + b)/2$, and these means are equal if and only if $a = b$.*

In our problem,

$$\frac{a + b}{2} \geq \sqrt{ab} \geq 2^3$$

and $(a + b)/2$ is smallest when equality holds, that is, when $a + b = 2 \cdot 2^3 = 16$.

Proof of italicized statement: Let x and y be any two real numbers. Then $(x - y)^2 = x^2 - 2xy + y^2 \geq 0$, and equality holds if and only if $x = y$. Hence $x^2 + y^2 \geq 2xy$, with equality if and only if $x = y$. Now set $x^2 = a$, $y^2 = b$; then the last inequality is equivalent to the celebrated $AM - GM$ inequality. $(a + b)/2 \geq \sqrt{ab}$, where equality holds if and only if $a = b$.

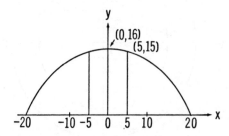

26. (B) Adopt a rectangular coordinate system (see figure), the x-axis being chosen along the span with the origin at its midpoint M. Then points A and B of the arch have coordinates $(-20, 0)$ and $(20, 0)$, and the vertex C of the parabola is at $(0, 16)$, so that its equation is $y = ax^2 + 16$. Since $B(20, 0)$ is on it, we have

$$0 = a \cdot 20^2 + 16, \quad a = -\tfrac{1}{25}. \quad \text{Thus } y = -\tfrac{1}{25}x^2 + 16.$$

Five feet from the center, $x = \pm 5$ and $y = -\tfrac{1}{25}(5)^2 + 16 = 15$ as stated in choice (B).

27. (E) The speed of the particle is the piecewise constant function

$$V_n = \text{speed of travelling the } n\text{-th mile}$$

$$= \frac{\text{distance}}{\text{time}} = \frac{1}{T_n},$$

where T_n is the number of hours needed to traverse the n-th mile. Since by hypothesis V_n is inversely proportional to $n - 1$, its reciprocal T_n is directly proportional to $n - 1$:

$$T_n = k(n - 1);$$

and when $n = 2$, $T_2 = k(2 - 1) = 2$. Hence $k = 2$ and the required time T_n is $2(n - 1)$.

Remark: Since the time needed to traverse the n-th mile clearly increases with n, choices (A) and (C) are immediately eliminated. Putting $n = 2$ then eliminates (B) and (D), leaving only (E).

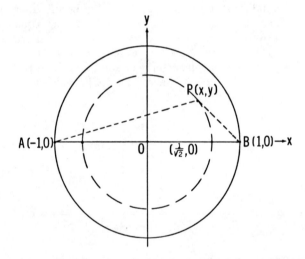

28. (E) Let the circle with radius 1 have equation $x^2 + y^2 = 1$ in a rectangular coordinate system (see figure). Without loss of generality, we may take $A(-1, 0)$ and $B(1, 0)$ as the ends of the given diameter. The condition to be satisfied by $P(x, y)$ is $AP^2 + PB^2 = 3$, or in terms of x and y

$$[(x + 1)^2 + y^2] + [(x - 1)^2 + y^2] = 3.$$

This simplifies to $2(x^2 + y^2) + 2 = 3$, or $x^2 + y^2 = \frac{1}{2}$. Hence the required points P are all points on the circle of radius $1/\sqrt{2}$, concentric with the given circle. The number of such points is infinite as stated in choice (E).

29. (C) If we divide the given expression for y by that for x, we obtain

$$\frac{y}{x} = \frac{t^{t/(t-1)}}{t^{1/(t-1)}} = t^{(t-1)/(t-1)} = t.$$

On the other hand,

$$y = t^{t/(t-1)} = (t^{1/(t-1)})^t = x^t,$$

and after substituting for t from our result above, we obtain

$$y = x^{y/x}, \quad \text{whence} \quad y^x = x^y.$$

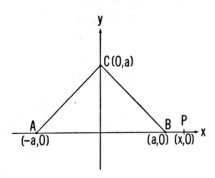

30. (D) Place the hypotenuse of $\triangle ABC$ on the x-axis with its center at the origin of the x, y-plane, and denote the coordinates of points A, B, C and P by $(-a, 0)$, $(a, 0)$, $(0, a)$ and $(x, 0)$, respectively (see figure). Then the expressions for s and CP^2 may be written

$$s = [x - (-a)]^2 + [x - a]^2 = 2(x^2 + a^2)$$

$$CP^2 = (0 - x)^2 + (a - 0)^2 = x^2 + a^2$$

so that $2CP^2 = s$ for all positions of P on the x-axis.

This problem can also be solved by expressing CP^2 by means of the law of cosines applied first to $\triangle CPA$, then to $\triangle CPB$.

Part 4

31. (D) Denote the given mapping from the xy- to the uv-plane by an arrow ($\rightarrow$), and the images of O, A, B, and C in the uv-plane by O', A', B', and C' (see figure). We have, by direct substitution,

$$O(0, 0) \rightarrow O'(0, 0), \qquad A(1, 0) \rightarrow A'(1, 0),$$

$$B(1, 1) \rightarrow B'(0, 2), \qquad C(0, 1) \rightarrow C'(-1, 0).$$

Segment OA from $(0, 0)$ to $(1, 0)$

$$\rightarrow \text{Segment } O'A' \text{ from } (0, 0) \text{ to } (1, 0).$$

Segment AB from $(1, 0)$ to $(1, 1)$

$$\rightarrow \begin{cases} \text{Parabolic arc } A'B' \text{ from } (1, 0) \text{ to } (0, 2) \\ \text{with equation } u = 1 - \frac{1}{4}v^2. \end{cases}$$

Segment BC from $(1, 1)$ to $(0, 1)$

$$\rightarrow \begin{cases} \text{Parabolic arc } B'C' \text{ from } (0, 2) \text{ to } (-1, 0) \\ \text{with equation } u = \frac{1}{4}v^2 - 1. \end{cases}$$

Segment CO from $(0, 1)$ to $(0, 0)$

$$\rightarrow \text{Segment } C'O' \text{ from } (-1, 0) \text{ to } (0, 0).$$

The transform (or image) of the square appears to be given by choice (D) of graphs.

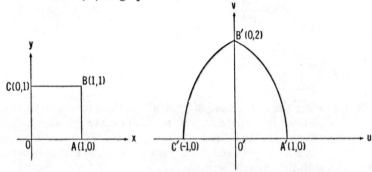

Comment: It is noteworthy that the transformation of this problem can be given by the single equation $w = z^2$, where $w = u + iv$ and $z = x + iy$ ($i = \sqrt{-1}$) are complex variables.

32. (C) If $u_n = a_0 + a_1 n + \cdots + a_k n^k$, then $u_1 = a_0 + a_1 + \cdots + a_k$. Hence the sum of the coefficients is $u_1 = 5$. It is of some interest, however, to find the actual polynomial expression for u_n.

We write the given recursion formula

$$u_{k+1} - u_k = 3 + 4(k - 1)$$

successively for $k = n - 1, n - 2, \cdots, 2, 1$ obtaining

$$u_n \quad - u_{n-1} = 3 + 4(n - 2)$$
$$u_{n-1} - u_{n-2} = 3 + 4(n - 3)$$
$$\cdots\cdots\cdots\cdots\cdots\cdots\cdots$$
$$\cdots\cdots\cdots\cdots\cdots\cdots\cdots$$
$$u_3 \quad - u_2 \quad = 3 + 4(1)$$
$$u_2 \quad - u_1 \quad = 3 + 4(0).$$

We add these $n - 1$ equations and observe that the left members form the "telescoping sum"

$$u_n - u_{n-1} + u_{n-1} - u_{n-2} + \cdots + u_2 - u_1 = u_n - u_1,$$

while the sum of the right members is

$$3(n - 1) + 4[1 + 2 + \cdots + n - 2]$$

$$= 3(n - 1) + 4\frac{(n - 2)(n - 1)}{2}\dagger$$

$$= (n - 1)[3 + 2(n - 2)]$$

$$= 2n^2 - 3n + 1.$$

Thus

$$u_n - u_1 = 2n^2 - 3n + 1,$$

and since $u_1 = 5$, the polynomial in n expressing u_n is

$$u_n = 2n^2 - 3n + 6;$$

the sum of its coefficients is $2 - 3 + 6 = 5$.

33. (A) Let a_1, a_2 denote the first terms, d_1 and d_2 the common differences of the arithmetic series with nth sums S_n and T_n, respectively. Then

$$S_n = n\left[a_1 + \frac{n - 1}{2}d_1\right], \qquad T_n = n\left[a_2 + \frac{n - 1}{2}d_2\right],$$

and

$$\frac{S_n}{T_n} = \frac{2a_1 + (n - 1)d_1}{2a_2 + (n - 1)d_2} = \frac{7n + 1}{4n + 27} \qquad \text{for all } n.$$

† See Comment following the solution of Problem 9, p. 114.

The eleventh terms of these series are $u_{11} = a_1 + 10d_1$ and $v_{11} = a_2 + 10d_2$, respectively, and their ratio is

$$\frac{u_{11}}{v_{11}} = \frac{a_1 + 10d_1}{a_2 + 10d_2} = \frac{2a_1 + 20d_1}{2a_2 + 20d_2}.$$

We note that the last expression is precisely S_n/T_n for $n = 21$, so that

$$\frac{u_{11}}{v_{11}} = \frac{7(21) + 1}{4(21) + 27} = \frac{148}{111} = \frac{4}{3}.$$

34. (B) The quotient in the division is a polynomial of degree 98 which we denote by $Q(x)$. Thus

$$x^{100} = Q(x)(x^2 - 3x + 2) + R.$$

Since the remainder R is of degree less than 2, we may denote it by $R(x) = ax + b$. Thus

$$x^{100} = Q(x)(x - 2)(x - 1) + (ax + b).$$

Setting first $x = 2$, then $x = 1$, we obtain

$$2^{100} = 2a + b \quad \text{and} \quad 1 = a + b.$$

Subtracting, we get $2^{100} - 1 = a$ from which $b = 1 - a = 1 - (2^{100} - 1) = 2 - 2^{100}$, so that

$$R(x) = ax + b = (2^{100} - 1)x + (2 - 2^{100})$$
$$= 2^{100}(x - 1) - (x - 2).$$

35. (B) The x-coordinates of the points of intersection of the graphs of $y = x^2 - 6$ and $y = m$ satisfy the equation $x^2 - 6 = m$ or $x^2 = 6 + m$. They are $x = \pm\sqrt{6 + m}$ which are real and not zero because $-6 < m < 6$, the left endpoint being $L(m) = -\sqrt{6 + m}$. Therefore

$$r = \frac{L(-m) - L(m)}{m} = \frac{-\sqrt{6 - m} - (-\sqrt{6 + m})}{m},$$

and, when the numerator is rationalized, this reduces to

$$r = \frac{2}{\sqrt{6 + m} + \sqrt{6 - m}}.$$

Hence when m approaches zero, the value of r approaches

$$\frac{2}{\sqrt{6} + \sqrt{6}} = \frac{2}{2\sqrt{6}} = \frac{1}{\sqrt{6}}.$$

1970 Solutions

Part 1

1. (E) Set $x = \sqrt{1 + \sqrt{1 + \sqrt{1}}}$. Since $\sqrt{1} = 1$, $x = \sqrt{1 + \sqrt{2}}$, $x^2 = 1 + \sqrt{2}$ and $x^4 = (x^2)^2 = 1 + 2\sqrt{2} + 2 = 3 + 2\sqrt{2}$.

2. (A) Let s, r, and p denote the side of the square, radius of the circle, and the common perimeter, respectively. Then $p = 4s = 2\pi r$, so $s = p/4$, and $r = p/(2\pi)$. Now let A_0 and A_s denote the areas of circle and square, respectively. Then the required ratio is

$$\frac{A_0}{A_s} = \frac{\pi r^2}{s^2} = \frac{\pi \left(\dfrac{p}{2\pi}\right)^2}{\left(\dfrac{p}{4}\right)^2} = \frac{4}{\pi}.$$

3. (C) To get y in terms of x, equate the expressions for 2^p obtained from the first and second given equations.

$$y - 1 = 2^{-p} = \frac{1}{x - 1} \quad \text{so that} \quad y = 1 + \frac{1}{x - 1} = \frac{x}{x - 1},$$

as stated in choice (C).

Comment: The given equations are parametric equations of the branch of a hyperbola which lies in the first quadrant with horizontal and vertical asymptotes $y = 1$ and $x = 1$. This is easily seen if we eliminate the parameter p by multiplying $x - 1 = 2^p$ by $y - 1 = 2^{-p}$ to obtain $(x - 1)(y - 1) = 2^0 = 1$ and keep in mind that $x > 1$ and $y > 1$ since $2^p > 0$ for all values of p.

4. (B) Three consecutive integers can always be expressed as $n - 1$, n, and $n + 1$, where n denotes the middle one. Thus each number of the set S is of the form

$$(n - 1)^2 + n^2 + (n + 1)^2 = 3n^2 + 2.$$

When n is even, $3n^2 + 2$ is divisible by 2, so choice (A) is false. We see that no member of S is divisible by 3, because the remainder in that division is always 2.

To eliminate choices (C) and (D), we show: (i) when n has a remainder of 1 upon division by 5, then $3n^2 + 2$ is divisible by 5; and (ii) when n has a remainder of 2 upon division by 7, then $3n^2 + 2$ is divisible by 7.

(i) If $n = 5m + 1$, $n^2 = 5^2m^2 + 2 \cdot 5m + 1$,

$$3n^2 + 2 = 3 \cdot 5^2m^2 + 6 \cdot 5m + 5 = 5[15m^2 + 6m + 1].$$

(ii) If $n = 7m + 2$, $n^2 = 7^2m^2 + 2 \cdot 2 \cdot 7m + 4$,

$$3n^2 + 2 = 3 \cdot 7^2m^2 + 12 \cdot 7m + 14 = 7[21m^2 + 12m + 2].$$

To show that (B) is correct, we must exhibit an n for which $3n^2 + 2$ is divisible by 11. This is so whenever n has the remainder 5 upon division by 11:

$$3(11m + 5)^2 + 2 = 3[11^2m^2 + 10 \cdot 11m + 5^2] + 2$$
$$= 11[33m^2 + 30m + 7].$$

Comment: The reader may wonder by what method we picked the proper integers n. Let us analyze the reasoning. Suppose we want to pick n so that $3n^2 + 2$ is divisible by the integer d. Let $n = kd + r$, $r < d$. Then

$$3n^2 + 2 = 3[kd + r]^2 + 2 = 3[k^2d^2 + 2kdr + r^2] + 2$$
$$= d[3k^2d + 6kr] + 3r^2 + 2.$$

This expression is divisible by d if and only if $3r^2 + 2$ is divisible by d. This seems not much easier to achieve than the original task, except that we need only test integers $r < d$:

r	0	1	2	3	4	5	$\cdots$
$3r^2 + 2$	2	5	14	29	50	77	$\cdots$

This shows that $3r^2 + 2$ is never divisible by 3; it is divisible by 5 when $r = 1$ or 4, and it is divisible by 11 when $r = 5$.

Such divisibility questions arise frequently and are most efficiently handled by congruences. We recommend that students acquaint themselves with a bit of modular arithmetic and congruences.

5. (D) Since $i = \sqrt{-1}$, $i^2 = -1$ and $i^4 = (i^2)^2 = (-1)^2 = 1$. Therefore

$$f(i) = \frac{i^4 + i^2}{1 + i} = \frac{1 - 1}{1 + i} = \frac{0}{1 + i} = 0,$$

because the denominator $1 + i \neq 0$, and the numerator is zero. [A complex number $a + bi$ is 0 if and only if $a = 0$ and $b = 0$].

6. (B) From the identity

$$x^2 + 8x = x^2 + 8x + 16 - 16 = (x + 4)^2 - 16,$$

we see that the given expression is least when the nonnegative expression $(x + 4)^2$ is zero. This occurs when $x = -4$. Then $x^2 + 8x$ is equal to $(-4)^2 + 8(-4) = -16$.

Comment: Real values of x are specified, because for complex values of x, the expression $x^2 + 8x$ may assume any value. We suggest that the reader show:

(a) If $x = u + iv$, $v \neq 0$, then $x^2 + 8x$ is real if and only if $u = -4$.

(b) Given any negative number N, it is possible to choose v so that $x^2 + 8x < N$ for $x = -4 + vi$.

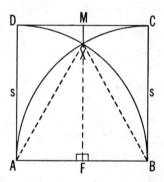

7. (E) The quarter circle arcs AXC and BXD intersecting at X have radii s, so that triangle ABX is equilateral. The line through X parallel to AD meets sides AB and DC at right angles in points F and M, respectively. The required distance from X to CD is $MX = s - XF = s - \frac{1}{2}s\sqrt{3}$ because XF is the altitude of the equilateral triangle ABX with side s and therefore has length $\frac{1}{2}s\sqrt{3}$. Hence $MX = \frac{1}{2}s(2 - \sqrt{3})$.

8. (B) The exponential form of the given equations is

$$8^a = 225, \qquad 2^b = 15.$$

Since $8 = 2^3$, and since $15^2 = 225$, we have

$$(2^3)^a = 2^{3a} = 225 = 2^{2b},$$

so that $3a = 2b$ and $a = 2b/3$.

9. (C) Since points P and Q divide AB in the ratios 2:3 and 3:4, respectively (see figure), they are $\frac{2}{5}$ and $\frac{3}{7}$ of the way from A to B. So

$$AP = \tfrac{2}{5}AB, \quad \text{and} \quad AQ = \tfrac{3}{7}AB.$$

Now $\quad PQ = AQ - AP = \dfrac{3}{7}AB - \dfrac{2}{5}AB = \dfrac{AB}{35}.$

We are told that $PQ = 2$, so $AB = 2 \cdot 35 = 70.$

10. (D) The repeating decimal F may be written as the sum of .4 and an infinite geometric series with common ratio .01:

$$F = .4818181\cdots = .4 + .0818181\cdots$$
$$= .4 + .081 + .00081 + \cdots$$
$$= .4 + .081(1 + .01 + .0001 + \cdots)$$
$$= .4 + .081 \cdot \frac{1}{1 - .01} = .4 + \frac{.081}{.99} = .4 + \frac{.9}{11} = \frac{53}{110}.$$

It is now seen that when the fraction F is written in lowest terms as above, the difference

$$\text{Denominator} - \text{Numerator} = 110 - 53 = 57.$$

Comment: The following evaluations of F, given without justification, involve the multiplication of a series by a constant and addition or subtraction of two series term by term.

$$
\begin{array}{ll}
100F = 48.1818\cdots & 10F = 4.8181\cdots \\
-F = -.4818\cdots & +F = .4818\cdots \\
\hline
99F = 47.7 & 11F = 5.2999\cdots = 5.3
\end{array}
$$

$$F = \frac{477}{990} = \frac{53}{110} \qquad\qquad F = \frac{53}{110}$$

Part 2

11. (E) Since two factors of the given cubic polynomial are known, the third linear factor, $2(x - c)$, can be determined as follows:

$$p(x) = 2x^3 - hx + k = 2\left[x^3 - \frac{h}{2}x + \frac{k}{2}\right]$$

$$= 2(x + 2)(x - 1)(x - c)$$

$$= 2[x^3 - (c - 1)x^2 - (c + 2)x + 2c].$$

Since the coefficient of x^2 is zero, $c = 1$, so that $p(x) = 2[x^3 - 3x + 2] = 2[x^3 - \frac{1}{2}hx + \frac{1}{2}k]$. Hence $h = 6$, $k = 4$, $|2h - 3k| = |12 - 12| = 0$.

OR

The factor theorem states that, if $x - r$ is a factor of a polynomial $p(x)$, then $p(r) = 0$. Thus

$$p(-2) = -16 + 2h + k = 0$$

$$p(1) = 2 - h + k = 0$$

and the unique solution of this linear system is $h = 6$, $k = 4$ as above.

Comment: Our first solution can be abbreviated by making use of the relations between the roots and the coefficients of a polynomial. Yet another solution would consist of dividing the given polynomial by the known factors $(x + 2)(x - 1)$ and setting the remainder equal to zero.

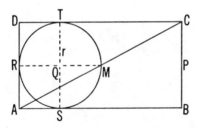

12. (C) Let Q denote the center of the given circle (see figure) and R, S, and T its points of tangency with sides AD, AB, and CD respectively. Points T, Q and S are collinear, and ST is a diameter of the circle and an altitude of the rectangle of length $2r$. RQ is parallel to sides AB and CD and midway between them. It therefore bisects every transversal, in particular the diagonal AC; so M, the midpoint of AC, lies on line RQ. Since M is also on the circle, RM is a diameter, and hence $RM = 2r$. Since $AD = 2AR$, we have $DC = 2RM = 4r$. The area of the rectangle is base·altitude = $4r·2r = 8r^2$.

13. **(D)** Using the definition $a * b = a^b$ of the operation $*$ which might be described as "exponentiation", we test each choice.

(A): $a * b = a^b$ while $b * a = b^a$; these are generally not equal so exponentiation is *not* commutative.

(B): $a * (b * c) = a * b^c = a^{b^c}$ and $(a * b) * c = a^b * c = (a^b)^c = a^{bc}$ are not always equal, so exponentiation is *not* associative.

(C): $(a * b^n) = a^{(b^n)}$ and $(a * n) * b = a^n * b = a^{nb}$ are not always equal, so choice (C) is wrong.

(D): $(a * b)^n = (a^b)^n = a^{bn}$ and $a * (bn) = a^{bn}$ are always equal, so (D) is a correct choice. Hence (E) is incorrect, so that (D) is the only correct choice.

14. **(A)** The roots of the given equation $x^2 + px + q = 0$, obtained by the quadratic formula, are

$$\tfrac{1}{2}(-p + \sqrt{p^2 - 4q}) \quad \text{and} \quad \tfrac{1}{2}(-p - \sqrt{p^2 - 4q}).$$

The difference of these roots is 1, so we must have $\sqrt{p^2 - 4q} = 1$, and hence $p^2 - 4q = 1$ so that $p = \sqrt{4q + 1}$. The negative value of the square root must be discarded because p and q are given as positive.

OR

Call the roots r and $r + 1$. Their sum is $-p = 2r + 1$ and their product is $q = r(r + 1)$. Thus

$$r = \frac{-p - 1}{2}, \qquad r + 1 = \frac{-p + 1}{2},$$

and

$$q = r(r + 1) = \frac{(-p - 1)(-p + 1)}{2 \cdot 2} = \frac{p^2 - 1}{4};$$

$p^2 = 4q + 1$, and $p = \sqrt{4q + 1}$.

15. **(E)** If points A and B have coordinates (x_A, y_A) and (x_B, y_B), then the coordinates (x_C, y_C) of a point C, which divides the segment AB so that $AC/AB = r$, satisfy

$$\frac{x_C - x_A}{x_B - x_A} = \frac{y_C - y_A}{y_B - y_A} = \frac{AC}{AB} = r,$$

as can be seen from the similar ring triangles with legs parallel to the coordinate axes and hypotenuses AC and AB. Solving these relations for x_C and y_C, we obtain

$$C: (x_C, y_C) = (x_A + r(x_B - x_A), \quad y_A + r(y_B - y_A)).$$

Now let $A = (-4, 5)$, $B = (5, -1)$. To find the trisection points P and Q, we use the ratios $1/3$ and $2/3$ respectively, obtaining

$$P = (x_P, y_P) = (-4 + \tfrac{1}{3}(9), \quad 5 + \tfrac{1}{3}(-6)) = (-1, 3)$$

$$Q = (x_Q, y_Q) = (-4 + \tfrac{2}{3}(9), \quad 5 + \tfrac{2}{3}(-6)) = (2, 1).$$

The difference of the y's divided by the difference of the x's of any pair of points on a line is constant;† for the pair (x, y) and the given point $(3, 4)$ this quotient is $(y - 4)/(x - 3)$. Equating this quotient to its values at the trisection points P and Q gives the equations of the required lines:

$$\frac{y - 4}{x - 3} = \frac{3 - 4}{-1 - 3} \quad \text{and} \quad \frac{y - 4}{x - 3} = \frac{1 - 4}{2 - 3},$$

which simplify to the equivalent equations

$$x - 4y + 13 = 0 \quad \text{and} \quad 3x - y - 5 = 0.$$

The first of these is that given in choice (E), and no other choice is equivalent to either of the two equations.

16. (C) Since $F(1) = F(2) = F(3) = 1$ we obtain $F(6)$ by computing $F(4)$ and then $F(5)$, using the given recurrence relation

$$F(n + 1) = \frac{F(n)F(n - 1) + 1}{F(n - 2)} \quad (n \geq 3).$$

Thus

$$F(4) = \frac{F(3)F(2) + 1}{F(1)} = \frac{1 \cdot 1 + 1}{1} = 2,$$

$$F(5) = \frac{F(4)F(3) + 1}{F(2)} = \frac{2 \cdot 1 + 1}{1} = 3,$$

and $\quad F(6) = \frac{F(5)F(4) + 1}{F(3)} = \frac{3 \cdot 2 + 1}{1} = 7.$

† In rectangular coordinates, this quantity is the slope of the line.

17. (E) We shall contradict choices (A), (B), (C), and (D) to establish the correctness of choice (E).

The given conditions $pr > qr$, $r > 0$ imply $p > q$, and $-p < -q$, contradicting (A); also if $p > 0$, then $1 > q/p$ contradicting (D). When $p > q \geq 0$, we have $q > -p$, which contradicts (B); and when p is positive, q negative, and $|q| > p$, then $-q > p > 0$ and $-q/p > 1$, which contradicts (C). Therefore (E) is the correct choice.

18. (A) Denote the required difference by d; d is positive, and

$$d^2 = (\sqrt{3 + 2\sqrt{2}})^2 - 2\sqrt{3 + 2\sqrt{2}}\sqrt{3 - 2\sqrt{2}} + (\sqrt{3 - 2\sqrt{2}})^2$$

$$= 3 + 2\sqrt{2} - 2\sqrt{3^2 - (2\sqrt{2})^2} + 3 - 2\sqrt{2}$$

$$= 6 - 2\sqrt{9 - 8} = 6 - 2 = 4.$$

Therefore $d = \sqrt{d^2} = \sqrt{4} = 2$.

Comment: An expression such as $\sqrt{3 + 2\sqrt{2}}$ may sometimes be simplified by assuming it to have the form $(x + y\sqrt{2})$ and determining x and y so that $(x + y\sqrt{2})^2 = 3 + 2\sqrt{2}$.† Thus $(x^2 + 2y^2) + 2xy\sqrt{2} = 3 + 2\sqrt{2}$ gives $x^2 + 2y^2 = 3$ and $2xy = 2$ or $y = 1/x$. Eliminating y, we have $x^2 + 2/x^2 = 3$, $x^4 - 3x^2 + 2 = 0$, $(x^2 - 1)(x^2 - 2) = 0$ so that $(x, y) = (1, 1)$ or $(-1, -1)$ or $(\sqrt{2}, \sqrt{2}/2)$ or $(-\sqrt{2}, -\sqrt{2}/2)$. The first of these gives the square root $x + y\sqrt{2} = 1 + \sqrt{2}$. The solution $(\sqrt{2}, \sqrt{2}/2)$ gives the same result. The negative of this obtained from $(-1, -1)$ or $(-\sqrt{2}, -\sqrt{2}/2)$ is discarded because the desired square root is positive. An analogous procedure yields $\sqrt{3 - 2\sqrt{2}} = \sqrt{2} - 1$, so that the difference $\sqrt{3 + 2\sqrt{2}} - \sqrt{3 - 2\sqrt{2}} = (1 + \sqrt{2}) - (\sqrt{2} - 1) = 2$ as before.

19. (C) Denote the first term of the geometric series by a; we are told that

$$a + ar + ar^2 + \cdots = a[1 + r + r^2 \cdots] = \frac{a}{1 - r} = 15,$$

so

$$a = 15(1 - r) = 15 - 15r.$$

† The student may try to find conditions on a, b, n so that $\sqrt{a + b\sqrt{n}}$ can be simplified to $x + y\sqrt{n}$.

The series of squares has sum

$$a^2 + a^2 r^2 + a^2 r^4 + \cdots = a^2 [1 + r^2 + r^4 \cdots]$$

$$= \frac{a^2}{1 - r^2} = \frac{a}{1 - r} \frac{a}{1 + r} = 45,$$

and when 15 is substituted for $a/(1 - r)$, we have $a/(1 + r) = 3$, so

$$a = 3(1 + r) = 3 + 3r.$$

Adding the equations

$$a = 15 - 15r \quad \text{and} \quad 5a = 15 + 15r$$

yields $6a = 30$, $a = 5$.

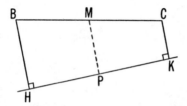

20. (A) In the figure, segment BC has midpoint M, and BH, CK are perpendicular to the line through HK, as required. Line MP, drawn perpendicular to HK, is parallel to BH and CK, bisects the transversal BC, and hence bisects every transversal including segment HK, of which it is therefore the perpendicular bisector. Hence M (and every point on MP) is equidistant from H and K, so that we always have $MH = MK$ as stated in choice (A). Choices (B) and (C) contradict (A) and hence are false. It is easy to see that (D) and (E) are false by constructing figures which satisfy the given conditions and violate (D) and (E).

Part 3

21. (B) The speedometer cable is so constructed that the mileage reading is proportional to the number N of revolutions the wheels make when traversing a distance D; the distance D is the product $2\pi r \cdot N$ of the circumference of the wheel and the number of revolutions it makes. Let r_1, r_2 be the radii of the wheels with regular and with snow tires, respectively, and let

N_1, N_2 be the numbers of revolutions the wheels make going and returning, respectively. Since the actual distances traversed are equal, we have $2\pi r_1 N_1 = 2\pi r_2 N_2$ so that

$$\frac{r_2}{r_1} = \frac{N_1}{N_2} = \frac{450}{440}.$$

Since $r_1 = 15$, we have $r_2 = 15 \cdot \frac{45}{44}$, and

$$r_2 - r_1 = 15(\tfrac{45}{44} - 1) = \tfrac{15}{44} \sim .34.$$

22. (A) Let S_m denote the sum of the first m positive integers. The formula for S_m is $S_m = \frac{1}{2}m(m+1)$ (see footnote on p. 114 for derivation) so that

$$S_{3n} - S_n = \tfrac{1}{2}3n(3n+1) - \tfrac{1}{2}n(n+1) = 4n^2 + n = 150.$$

Thus

$$4n^2 + n - 150 = (n-6)(4n+25) = 0, \quad n = 6 \text{ or } -\tfrac{25}{4}.$$

Since n must be a positive integer, $n = 6$, $4n = 24$, and

$$S_{4n} = S_{24} = \tfrac{1}{2}24(24+1) = 12 \cdot 25 = 300.$$

Remark: $S_m = \frac{1}{2}m(m+1)$ has the property that $8S_m + 1$ is a perfect square:

$$8S_m + 1 = 4m(m+1) + 1 = 4m^2 + 4m + 1 = (2m+1)^2.$$

Among the numbers in choices (A) through (E), only 300 has that property; so we could have eliminated (B) through (E) without using the hypothesis $S_{3n} - S_n = 150$, and without calculating n.

23. (D) In the base 10,

$$10! = 1 \cdot 2 \cdot 3 \cdot 4 \cdot 5 \cdot 6 \cdot 7 \cdot 8 \cdot 9 \cdot 10 = 2^8 \cdot 3^4 \cdot 5^2 \cdot 7$$

$$= 12^4 \cdot 5^2 \cdot 7$$

and $5^2 \cdot 7 = 175 = 1 \cdot 12^2 + 2 \cdot 12 + 7$. Thus

$$[5^2 \cdot 7 \cdot 12^4]_{10} = [127 \cdot 10^4]_{12}$$

$$= 1,270,000_{12}.$$

This number ends with exactly 4 zeros.

24. (B) Let s denote the length of a side of the hexagon. Since its perimeter $6s$ is the same as that of the triangle, each side of the triangle has length $2s$. Now the triangle of given area 2

can be cut into four congruent equilateral triangles having sides of length s, while the hexagon can be cut up into six such triangles (see figure). Therefore

$$\frac{\text{Area of hexagon}}{\text{Area of triangle}} = \frac{\text{Area of hexagon}}{2} = \frac{6}{4} = \frac{3}{2},$$

so Area of hexagon $= \frac{3}{2}\cdot 2 = 3$.

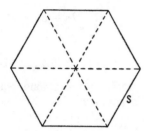
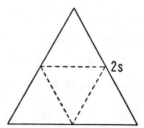

25. (E) The statement of the problem defines the function

$$[x] = \text{greatest integer} \le x.$$

To calculate postage, we need the slightly different function

$$L(x) = \text{least integer} \ge x$$

corresponding to "every ounce or portion thereof". Since our answer involves the function $[x]$, we need to express $L(x)$ in terms of $[x]$ and claim that $L(x) = -[-x]$. To prove this claim, write

$$x = n + \alpha, \quad n \text{ an integer}, \quad 0 \le \alpha < 1.$$

Then

$$[x] = n, \qquad L(x) = \begin{cases} n & \text{if } \alpha = 0 \\ n+1 & \text{if } \alpha \ne 0. \end{cases}$$

Now $-x = -n - \alpha$, so

$$[-x] = \begin{cases} -n & \text{if } \alpha = 0 \\ -n-1 & \text{if } \alpha \ne 0 \end{cases} \quad \text{and} \quad -[-x] = \begin{cases} n & \text{if } \alpha = 0 \\ n+1 & \text{if } \alpha \ne 0, \end{cases}$$

and the last description agrees precisely with that of $L(x)$. Thus the required expression for the postage is

$$6L(W) = -6[-W].$$

Questions: (i) Does $[x] = -L(-x)$? (ii) For what values of x does $L(x) = [x]$ hold? (iii) How many different values can the function $L(x) - [x]$ assume? (iv) What special names are given to $[\log_{10} x]$ and $\log_{10} x - [\log_{10} x]$?

26. (B) The two lines which are the graph of the first equation intersect at the point $(2, 3)$ obtained by solving the simultaneous equations

I: $x + y - 5 = 0$ and   II: $2x - 3y + 5 = 0.$

Similarly, the lines

III: $x - y + 1 = 0$ and   IV: $3x + 2y - 12 = 0,$

whose graphs constitute the graph of the second equation, have the same point $(2, 3)$ in common. Since all four lines have different slopes (in fact, I and III are perpendicular, II and IV are perpendicular), $(2, 3)$ is the only point common to both graphs.

Question: If the factor $(x - y + 1)$ in the second equation is replaced by $(x + y + 1)$, then the graphs of the two equations have exactly two points in common. Can you explain this?

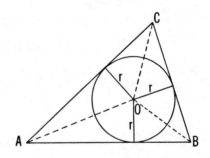

27. (A) Let the given triangle be ABC with perimeter $p = AB + BC + CA$, and denote the center and radius of its inscribed circle by O and r; see figure. Then the area of $\triangle ABC$ is the sum of the areas of $\triangle AOB$, $\triangle BOC$, and $\triangle COA$, whose bases are AB, BC and CA respectively, and whose altitudes have length r. Therefore

$$\text{Area of } \triangle ABC = \tfrac{1}{2}rAB + \tfrac{1}{2}rBC + \tfrac{1}{2}rCA$$

$$= \tfrac{1}{2}r(AB + BC + CA) = \tfrac{1}{2}rp,$$

which is given to be equal to the perimeter p of $\triangle ABC$: $\tfrac{1}{2}rp = p$. Hence $r = 2$.

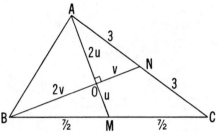

28. (A) Let the perpendicular medians from A and B intersect each other at O and their opposite sides at their midpoints M and N respectively, so $AN = 3$ and $BM = \frac{7}{2}$ (see figure). Let segments AO and BO have lengths $2u$ and $2v$, respectively, so that OM and ON have lengths u and v. Then the Pythagorean Theorem applied to right triangles AON and BOM yields respectively

$$4u^2 + v^2 = 3^2 = 9 \quad \text{and} \quad u^2 + 4v^2 = (7/2)^2 = 49/4.$$

Four-fifths of the sum of these two equations gives $4u^2 + 4v^2 = 17$ which is equal to the square of the hypotenuse AB of right triangle AOB. Hence the length of AB is $\sqrt{17}$.

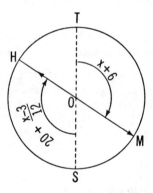

29. (D) Let x denote the number of minutes after 10 o'clock now. Let M and H (see figure) be the points on the dial to which the minute and hour hands point 6 minutes hence and 3 minutes ago, respectively. If O is the center and T and S the twelve and six marks on the dial, respectively, then (measured in minute spaces) the angles $TOM = x + 6$ and $SOH = 20 + (x - 3)/12$. But these are equal vertical angles because TS and HM are both straight lines through O. The equation

$$x + 6 = 20 + (x - 3)/12 \qquad \text{yields} \qquad x = 15,$$

so that the time now is 10:15.

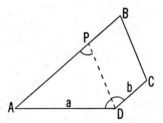

30. (E) Let the bisector of $\measuredangle D$ intersect AB at P (see figure). Then the alternate interior angles APD and PDC as well as $\measuredangle ADP$ are equal to angle B, so that $\triangle APD$ is isosceles with equal angles at P and D. This makes $AP = AD = a$. Since $PBCD$ is a parallelogram, we have $PB = DC = b$; so $AB = AP + PB = a + b$.

Part 4

31. (B) Since the largest possible digit in base 10 is 9, the sum $d_1 + d_2 + d_3 + d_4 + d_5$ of the five digits can be at most 45. The given sum, 43, is two less and can come about in the following ways:

(i) One of the digits is 7 (2 less than 9), all others are 9; 7 may appear in 5 possible places, 79999, 97999, 99799, 99979, 99997.

(ii) Two of the digits are 8 (each one less than 9), the other three are 9. This can happen in $5 \cdot 4/2 = 10$ ways, 88999, 89899, 89989, 89998, 98899, 98989, 98998, 99889, 99898, 99988. Next we recall that a number is divisible by 11 if and only if the alternating sum of its digits $d_1 - d_2 + d_3 - d_4 + d_5$ is divisible by 11.† We find that exactly three of the 15 numbers, namely 97999, 99979, 98989, are divisible by 11, so the required probability is $\frac{3}{15} = \frac{1}{5}$.

† This fact is based on an important property of integers: Let R be the remainder when a sum $N_1 + N_2 + \cdots + N_k$ is divided by D; i.e.

$$N_1 + N_2 + \cdots + N_k = QD + R,$$

and let R_i be the remainder when N_i is divided by D; i.e.

$$N_i = Q_i D + R_i, \qquad i = 1, 2, \cdots, k.$$

Then R is equal to the remainder in the division of the sum $R_1 + R_2 + \cdots + R_k$ by D:

$$R_1 + R_2 + \cdots + R_k = PD + R.$$

Comment: Divisibility by 11 (or any other prime integer) is handled more simply by congruence modulo 11 (or the other prime). Here 10^k is congruent to 1 or -1 according as k is even or odd. Hence we see that the congruence

$$d_1 + 10d_2 + 10^2d_3 + 10^3d_4 + 10^4d_5 \equiv 0 \pmod{11}$$

reduces to $d_1 - d_2 + d_3 - d_4 + d_5 \equiv 0 \pmod{11}$ (this is the divisibility criterion found before). We are given the sum

$$d_1 + d_2 + d_3 + d_4 + d_5 = 43 \equiv -1 \pmod{11}.$$

When the preceding congruence is subtracted from this, we obtain $2d_2 + 2d_4 \equiv -1 \equiv 10 \pmod{11}$, so that $d_2 + d_4 \equiv 5 \pmod{11}$. Hence $d_2 + d_4 = 16$ and d_2 and d_4 must be either 8 and 8, or 7 and 9, or 9 and 7, respectively, because 7, 8, 9 are the only allowable digits. The resulting numbers are 98989, 97999, 99979 as before.

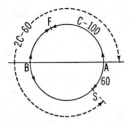

32. (C) Let $2C$ denote the number of yards in the circumference of the track, A and B the starting points, F and S the positions of first and second meeting. On first meeting, the distances travelled by A and B, respectively, are $(C - 100)$ and 100, and on second meeting, $(2C - 60)$ and $(C + 60)$. Since each travels at uniform speed, the ratio of their distances is the same for every time interval. In particular, at F and S,

$$\frac{C - 100}{100} = \frac{2C - 60}{C + 60} \quad \text{so} \quad C = 240.$$

Therefore the circumference $2C$ is 480 yards.

Now each term of our five digit number $d_1 + 10d_2 + \cdots + 10^4d_5$ is of the form $10^n d_{n+1}$, and since $10 = 11 - 1$, we see from the binomial expansion that

$$10^n d_{n+1} = (11 - 1)^n d_{n+1} = s \cdot 11 + (-1)^n d_{n+1},$$

$n = 1, 2, \cdots, 5$, so that the sum of the remainders R_i is

$$(-1)^0 d_1 + (-1)^1 d_2 + (-1)^2 d_3 + (-1)^3 d_4 + (-1)^4 d_5 = d_1 - d_2 + d_3 - d_4 + d_5$$

$$= P \cdot 11 + R.$$

33. (A) Omitting 10,000 and including 0 momentarily, reduces the required sum of digits by 1. Multiplying each number in this new sequence 0, 1, 2, 3, $\cdots$, 9999 by 10^{-4} (or any other power of ten) does not change the sum of the digits, but gives the sequence .0000, .0001, .0002, $\cdots$, .9999 of all 10,000 four place decimal fractions. Each of the ten digits 0, 1, 2, $\cdots$, 9 appears the same number of times, which is $10000/10 = 1000$ times, in each of the four decimal places of the 10,000 decimal fractions, so each digit occurs $4 \cdot 1000 = 4000$ times in all. The sum of all digits is therefore

$$4000(0 + 1 + 2 + \cdots + 9) = 4000(45) = 180,000.$$

Now adding the 1 by which we reduced the sum of digits, when we included 0 and omitted 10,000 momentarily, brings the total to 180,001.

34. (C) If three integers a, b, and c have the same remainder r upon division by an integer d, then

$$a = \alpha d + r, \quad b = \beta d + r, \quad \text{and} \quad c = \gamma d + r,$$

where α, β, γ are the quotients. The differences

$$a - b = (\alpha - \beta)d, \qquad a - c = (\alpha - \gamma)d,$$

and
$$b - c = (\beta - \gamma)d$$

are exactly divisible by d. Moreover, since $(a - b) - (a - c) + (b - c) = 0$, any common divisor d of two of the differences is a divisor of the third. Hence the G.C.D. (greatest common divisor) of any pair of the differences is the greatest integer leaving the same remainder when divided into all three of the original numbers a, b, and c.

In the present problem, we seek the G.C.D. of the two differences

$$13,903 - 13,511 = 392 = 7^2 \cdot 2^3$$

and
$$14,589 - 13,903 = 686 = 7^3 \cdot 2$$

which, by inspection, is $7^2 \cdot 2 = 98$.

Comment: The Euclidean Algorithm furnishes an automatic arithmetic process for finding the G.C.D. of any two integers. For an elementary discussion and proof, see for example *Continued Fractions* by C. D. Olds, vol. 9 in this NML series, Random House/Singer (1963), p. 17, or *College Algebra* by Fine.

35. (D) Let X denote the amount of the annual pension and y the number of years of service. Then with constant of proportionality k, the statement of the problem yields the following three equations (and their squares below them):

$$X = k\sqrt{y}, \qquad X + p = k\sqrt{y+a}, \qquad X + q = k\sqrt{y+b}$$

$$X^2 = k^2y, \qquad (X+p)^2 = k^2(y+a), \qquad (X+q)^2 = k^2(y+b).$$

Replacing k^2y by X^2 in the last two equations and simplifying gives

$$2pX + p^2 = k^2a \qquad \text{and} \qquad 2qX + q^2 = k^2b.$$

We divide the first by the second equation and solve for X:

$$\frac{2pX + p^2}{2qX + q^2} = \frac{a}{b}, \qquad X = \frac{aq^2 - bp^2}{2(bp - aq)}.$$

We note that X is not defined when $bp = aq$; but then the next to the last pair of equations multiplied by q and p, respectively, give

$$2pqX + p^2q = k^2aq \qquad \text{and} \qquad 2pqX + q^2p = k^2bp,$$

and subtracting one from the other,

$$pq(p - q) = k^2(aq - bp) = k^2 \cdot 0 = 0.$$

Hence either $pq = 0$, or $p - q = 0$. In the first case, p or q is zero so that a or b is zero and the second or the third of our original 3 equations would be identical with the first. In the second case, $p = q$ which would require $a = b$ contrary to hypothesis.

1971 Solutions

Part 1

1. (B) The factors 2 and 5 in the given number are the factors of 10, and it is easy to count digits of a number when it is written as a multiple of a power of 10. Accordingly, we write

$$N = 2^{12} \cdot 5^8 = 2^4 \cdot 2^8 \cdot 5^8 = 2^4 (10)^8 = 16 \cdot 10^8 = 1,600,000,000$$

and see that N is a 10 digit number.

Question: Can you see that, if the base of the number system were twenty instead of ten, the number of digits in N would be eight? How about base 50?

2. (D) The number x of bricks laid is jointly proportional to the number y of men and the number z of days: $x = kyz$. The given information says that when $y = b$ and $z = c$, then $x = f$. Hence $f = kbc$, which determines the constant of proportionality $k = f/(bc)$, so that $x = fyz/(bc)$. When the number x of bricks is b and the number y of men is c, we get $b = fz/b$, from which the number of days is $z = b^2/f$.

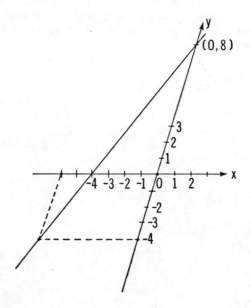

3. (E) The quotient of the difference of the y and x coordinates of any two points on a line in the xy-plane remains constant. Equating these quotients for the two pairs of points $(x, -4)$, $(-4, 0)$ and $(-4, 0)$, $(0, 8)$ gives the equation

$$\frac{0-(-4)}{-4-x} = \frac{8-0}{0-(-4)}, \quad \text{or} \quad -\frac{4}{x+4} = 2$$

which yields $x = -6$.

Comment: When a rectangular xy-coordinate system is chosen (as is usually the case), the above quotient is called the slope of the line. The result however, is the same for any Cartesian coordinate system.

4. (A) Let P, r and t denote the principal, rate of simple interest, and time in years, respectively. Then the given information may be written

$$255.31 = P + Prt = P(1 + rt),$$

where Prt is the interest credited. Since $r = .05$, and $t = 1/6$,

$$1 + rt = 1 + \frac{5}{600} = \frac{605}{600}, \quad \text{so} \quad P = \frac{255.31}{1 + rt} = \frac{255.31 \cdot 600}{605},$$

and

$$rtP = \frac{5}{600}P = \frac{255.31 \cdot 5}{605} = \frac{255.31}{121} = 2.11,$$

so that the number of cents in the credited interest is 11.

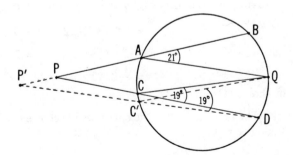

5. (C) We have $\angle P = \frac{1}{2}(\widehat{BD} - \widehat{AC})$ and $\angle Q = \frac{1}{2}\widehat{AC}$. Hence $\angle P + \angle Q = \frac{1}{2}\widehat{BD} = \frac{1}{2}(42° + 38°) = 40°$. Note that the given data force the sum of the measures of angles P and Q to be constant while the measure of each may vary (see figure).

6. (E) Since by definition $a * b = 2ab$ over S,

    (A) $*$ is commutative because $a * b = 2ab = 2ba = b * a$;

    (B) $*$ is associated because

$$a * (b * c) = 2a(b * c) = 2a(2bc) = 4abc$$

which equals

$$(a * b) * c = 2(a * b)c = 2(2ab)c = 4abc;$$

    (C) $\frac{1}{2}$ is a left identity because $\frac{1}{2} * a = 2(\frac{1}{2}a) = a$ for any a in S, and moreover, $\frac{1}{2}$ is a right identity because $*$ is commutative.

    (D) Every element a of S has a left inverse $l = 1/4a$ because

$$l * a = \frac{1}{4a} * a = \frac{2}{4a} \cdot a = \frac{1}{2} = \text{identity.}$$

Moreover, since $*$ is commutative, $a * 1/4a$ is also the identity $\frac{1}{2}$, so $1/4a$ is also a right inverse of a.

    (E) The element $1/2a$ is not an inverse for the element a of S because

$$\frac{1}{2a} * a = a * \frac{1}{2a} = 2\left(\frac{1}{2a}\right) a = 1 \neq \frac{1}{2}.$$

Therefore (E) is the only incorrect statement.

7. (C) The given expression

$$2^{-(2k+1)} - 2^{-(2k-1)} + 2^{-2k} = 2^{-2k-1} - 2^{-2k+1} + 2^{-2k}$$
$$= 2^{-2k} \cdot 2^{-1} - 2^{-2k} \cdot 2 + 2^{-2k} \cdot 1$$
$$= 2^{-2k}(\tfrac{1}{2} - 2 + 1) = 2^{-2k}(-\tfrac{1}{2})$$
$$= -2^{-2k} 2^{-1} = -2^{-2k-1} = -2^{-(2k+1)}$$

or choice (C).

Remark: The fact that (C) is the only possible choice can be seen by taking $k = 0$.

8. (B) The given conditional inequality is equivalent to

$$6x^2 + 5x - 4 < 0, \qquad (3x + 4)(2x - 1) < 0.$$

Now when $3x + 4 < 0$, then

$$x < \frac{-4}{3}, \quad \text{and} \quad 2x - 1 < \frac{-8}{3} - 1 < 0,$$

so that both factors are negative. When $2x - 1 > 0$, then

$$x > \tfrac{1}{2}, \quad \text{and} \quad 3x + 4 > 3 \cdot (\tfrac{1}{2}) + 4 > 0,$$

so that both factors are positive. In either case, the product is positive. But when

$$-\tfrac{4}{3} < x < \tfrac{1}{2}, \quad \text{then} \quad 3x + 4 > 0 \quad \text{and} \quad 2x - 1 < 0;$$

the factors have opposite signs, so their product, $6x^2 + 5x - 4$, is negative. Since either $x = -\tfrac{4}{3}$ or $x = \tfrac{1}{2}$ make the product 0, the solution set consists of all values of x such that $-\tfrac{4}{3} < x < \tfrac{1}{2}$ as stated in choice (B).

OR

One may consider the graph of the function $y = 6x^2 + 5x - 4$ which is a parabola lying below the x-axis when $-\tfrac{4}{3} < x < \tfrac{1}{2}$ (see figure). Thus y is negative when x is in the interval $-\tfrac{4}{3} < x < \tfrac{1}{2}$, zero at the endpoints of that interval, and otherwise y is positive.

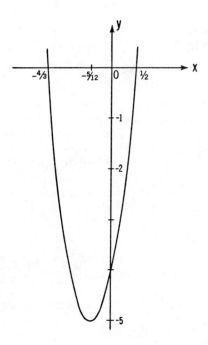

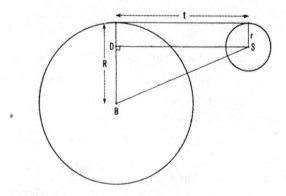

9. (D) Consider two circles with centers B and S and radii R and r, respectively ($r < R$, see figure) extending to the points of contact of their common external tangent of length t. Draw segment SD from S perpendicular to radius R at D. Then BDS is a right triangle with hypotenuse equal to the distance BS between the centers, so that

$$BS^2 = SD^2 + BD^2 = t^2 + (R - r)^2$$

by the Pythagorean Theorem.

In the present problem t, R, and r are given as 24, 14, and 4 inches, respectively, so that

$$BS^2 = 24^2 + 10^2 = 2^2[12^2 + 5^2] = 2^2 \cdot 13^2, \quad BS = 2 \cdot 13 = 26.$$

10. (E) The situation in this problem (and many others) may be defined in terms of the presence or absence of two mutually exclusive properties. Thus, let p and q denote the sets of all brown-eyed girls and all brunettes, respectively, so that $\sim p$ (not p) and $\sim q$ (not q) are the sets of blue-eyed girls and blondes, respectively. Then there are exactly four basic intersections: brown-eyed brunettes, brown-eyed blondes, blue-eyed brunettes and blue-eyed blondes, denoted by the products

$$pq, \quad p(\sim q), \quad (\sim p)q, \quad (\sim p)(\sim q),$$

with x, y, z, w girls, respectively, in each. Since the basic sets are disjoint, the given information is expressed by the four equations

$$x + y + z + w = 50, \quad w = 14,$$
$$x + z = 31, \quad x + y = 18.$$

The required number x of brown-eyed brunettes is obtained by adding the last three equations and subtracting the first:

$$x = [w + (x + z) + (x + y)] - (x + y + z + w)$$

$$= (14 + 31 + 18) - 50 = 13.$$

Remark: There are many other ways of solving the linear system of 4 equations in x, y, z, w for x.

OR

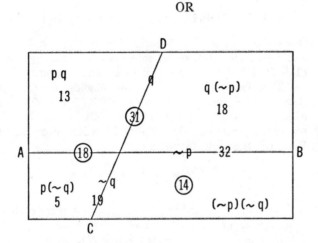

We may use a Venn Diagram (see figure) to represent sets schematically by regions. Thus the whole rectangle represents all 50 girls with p the 18 brown-eyed and hence $\sim p$ the 32 blue-eyed represented by the regions to the left and right of line CD, and with q the 31 brunettes and hence $\sim q$ the 19 blondes represented above and below the line AB respectively. Given numbers are circled, and designations on line segments apply to the regions on both sides of them. Now the given number 14 of blue-eyed blondes in set $(\sim p)(\sim q)$ may be subtracted from the total of 19 blondes in set $(\sim q)$ to give 5 brown-eyed blondes in set $p(\sim q)$ which may be subtracted from the given 18 brown-eyed girls in set p to give the required 13 brown-eyed brunettes in the set pq as stated in choice (E).

Remark: We might equally well have proceeded counterclockwise in the Venn Diagram, subtracting 14 from 32 to get 18, which subtracted from 31 gives the required 13 in set pq as before.

Part 2

11. (D) First, bases a and b must both exceed 7 to make either representation 47 or 74 possible. We are told that the numbers

$$(47)_a = 4a + 7 \quad \text{and} \quad (74)_b = 7b + 4$$

are equal, so that

$$4a + 7 = 7b + 4 \quad \text{or equivalently} \quad 7b - 4a = 3.$$

One solution of the last equation is evidently $(a, b) = (1, 1)$ and hence all solutions in integers are given by $(a, b) = (1 + 7t, 1 + 4t)$, where t may be any integer.† The value $t = 2$ makes both a and b the smallest solutions greater than 7. Therefore $(a, b) = (1 + 7 \cdot 2, 1 + 4 \cdot 2) = (15, 9)$ makes the sum $a + b = 15 + 9 = 24$ least. The Roman numeral representing the number 24 is XXIV or choice (D). We check the result by noticing that

$$(67)_{10} = (47)_{15} = (74)_9.$$

12. (B) The number N, usually called the modulus of the system, is an exact divisor of the difference of any two congruent numbers. For if a and b are congruent (mod N), then they have the same remainder upon division by N so that $a = kN + r$, $b = lN + r$, and $a - b = (k - l)N$. In this problem, $N = 7$ because the differences $90 - 69 = 21$, $125 - 90 = 35$, and $125 - 69 = 56$ are divisible by 7 and have no other common divisor. Now $81 = 11 \cdot 7 + 4$ so that 81 is congruent to 4 modulo 7.

Remark: "Congruence modulo N" is an equivalence relation and divides the set of all integers into N disjoint equivalence classes corresponding to the N possible remainders 0, 1, 2, $\cdots$, $N - 1$ in division by N. For $N = 7$ (or any other prime modulus), the system is a "field" in which division by any integer not congruent to zero is always possible. The quotient a/b is defined as the solution x of the congruence $bx \equiv a(\mathrm{mod}\, N)$. For example, the quotient 81/125 when $N = 7$, is given by $125x \equiv 81(\mathrm{mod}\, 7)$, which reduces to $x \equiv 3(\mathrm{mod}\, 7)$.

† Solutions in integers of equations of the form $mx - ny = c$, where m, n and c are integers, are fully discussed in *Continued Fractions*, C. D. Olds, vol. 9 in this NML series, pp. 36–42.

13. (E) We write 1.0025 as the sum $1 + .0025$ and observe that the terms in the binomial expansion

$$(1 + .0025)^{10} = 1 + 10(.0025) + \frac{10 \cdot 9}{1 \cdot 2}(.0025)^2$$

$$+ \frac{10 \cdot 9 \cdot 8}{1 \cdot 2 \cdot 3}(.0025)^3 + \cdots$$

$$= 1 + .025 + .0028125 + .000001875 + R$$

$$= 1.025283125 + R$$

decrease so rapidly that only the first four terms affect the first five decimal places of the sum. To see this, we shall estimate

$$R = \sum_{k=4}^{10} \binom{10}{k} b^k.\dagger$$

Here b stands for

$$.0025 = \frac{1}{2^2 \cdot 10^2} < 1, \text{ and } \binom{10}{k} = \frac{10!}{k!(10-k)!}$$

denotes the coefficient of x^k in the expansion of $(1 + x)^{10}$. We note first that

$$\sum_{k=0}^{10} \binom{10}{k} = (1+1)^{10} = 2^{10},$$

and that all terms in the sum are positive.

$$R = \sum_{k=4}^{10} \binom{10}{k} b^k < b^4 \sum_{k=4}^{10} \binom{10}{k} \quad \text{(because } 0 < b < 1\text{)}$$

$$< b^4 \sum_{k=0}^{10} \binom{10}{k} \quad \text{(we have augmented the sum)}$$

$$= b^4 \cdot 2^{10}.$$

Thus

$$R < \frac{2^{10}}{2^8 10^8} = \frac{2^2}{10^8} = .00000004.$$

Hence $(1.0025)^{10} \approx 1.02528$ correct to 5 decimal places. The fifth decimal place has the digit 8.

† The symbol $\sum_{i=1}^{n} a_i$ stands for $a_1 + a_2 + \cdots + a_n$.

Remark: In order to evaluate a power x^n, it is often convenient to write $x = A + B$ with B small compared to A and to approximate

$$x^n = (A + B)^n = A^n \left(1 + \frac{B}{A}\right)^n$$

by the product

$$A^n \cdot \left\{\text{first few terms of the binomial expansion of } \left(1 + \frac{B}{A}\right)^n\right\}.$$

The error committed can easily be estimated.

14. (C) We can see that 63 and 65 are both divisors by straightforward factoring. Thus

$$2^{48} - 1 = (2^{24} - 1)(2^{24} + 1) = (2^{12} - 1)(2^{12} + 1)(2^{24} + 1)$$
$$= (2^6 - 1)(2^6 + 1)(2^{12} + 1)(2^{24} + 1)$$
$$= 63 \cdot 65(2^{12} + 1)(2^{24} + 1).$$

15. (B) Let u and h inches, respectively, be the length of the bottom and depth of the water when the bottom is level. The volume of the water which is then in the form of a rectangular parallelepiped, is $10hu$ cubic inches. When the aquarium is tilted, the volume of the water which is now in the form of a prism with altitude 10 inches and right triangular base, is

$$10 \cdot \tfrac{1}{2} \cdot 8(\tfrac{3}{4}u) = 10 \cdot 3u$$

cubic inches. Equating these two expressions for the volume of water gives

$$10uh = 10 \cdot 3u, \quad \text{so} \quad h = 3.$$

Thus the depth of the water when the bottom is level is 3 inches.

16. (A) Let the 35 scores be denoted by $x_1, x_2, x_3, \cdots, x_{35}$ and their average by $\bar{x}$. Then the average A of all 36 numbers is

$$A = \tfrac{1}{36}(35\bar{x} + \bar{x}) = \tfrac{1}{36}(36\bar{x}) = \bar{x},$$

and the required ratio $A/\bar{x} = 1$.

17. (E) Denote the center of the circular disk by O (see figure), and let the radii $r_1, r_2, r_3, \cdots, r_{2n}$ cut off equal arcs $a_1, a_2, a_3, \cdots, a_{2n}$, proceeding counterclockwise from r_{2n} which is taken, without loss of generality, as extending to the right. Let P and Q be

any interior points of radii r_1 and r_n, respectively. Then POQ is a triangle whose base PQ is a segment of the secant ST cutting arc a_1 in S and arc a_{n+1} in T. Segment PQ divides each of the $(n-1)$ sectors subtended by arcs $a_2, a_3, \cdots, a_n$ into two parts. Segments PS and QT cut the two bordering sectors subtended by arcs a_1 and a_{n+1}, dividing each into two parts. In all, $(n-1) + 2 = n+1$ of the $2n$ disjoint sectors are cut in two, giving a total of $2n + (n+1) = 3n+1$ regions. This is the maximum number because the $2n$ radii constitute n lines which can be cut by one secant (line) in at most n points.

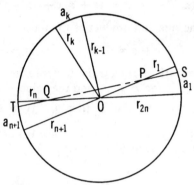

18. (D) Let v denote the boat's rate in still water in miles per hour. Then the time (distance/rate) for 4 miles downstream and return, which totals 1 hour, gives the equation

$$\frac{4}{v+3} + \frac{4}{v-3} = 1,$$

since the stream's rate is 3 miles per hour. Multiplying both sides by $(v+3)(v-3)$ and simplifying yields $v^2 - 8v - 9 = (v-9)(v+1) = 0$, $v = 9$ ($v = -1$ must be rejected). Now, since the rate downstream is $v+3 = 9+3 = 12$ and the rate upstream is $v-3 = 9-3 = 6$, their ratio is two to one.

19. (C) The values of x at the points of intersection of the line and ellipse are the solutions of the quadratic equation

$$x^2 + 4(mx+1)^2 = 1 \quad \text{or} \quad (1+4m^2)x^2 + 8mx + 3 = 0$$

obtained by substituting $mx+1$ for y in the equation $x^2 + 4y^2 = 1$ of the ellipse. The condition that there be exactly one intersection implies that the quadratic equation have only one root, which means that its discriminant is zero. That is,

$$(8m)^2 - 4(1 + 4m^2) \cdot 3 = 0.$$

This reduces to $m^2 = \frac{3}{4}$ or choice (C). Incidentally, the condition for 2 intersections or none is that the discriminant be positive or negative, respectively.

20. (E) The given equation is equivalent to $x^2 + 2hx - 3 = 0$. Denoting its roots by r and s, we have

$$(x - r)(x - s) = x^2 - (r + s)x + rs = x^2 + 2hx - 3 = 0,$$

so that the sum and product of the roots are

$$r + s = -2h, \qquad rs = -3,$$

respectively. If we square the first relation and substitute the given value 10 for $r^2 + s^2$, and the value -3 for rs, we obtain

$$(r + s)^2 = r^2 + 2rs + s^2 = 10 + 2(-3) = 4 = (-2h)^2 = 4h^2,$$

so that $h^2 = 1$ and $|h| = 1$. Since 1 is not offered in choices (A)–(D), statement (E) is correct.

Part 3

21. (C) Since, for any base $b \neq 0$, $\log_b N = 0$ only if $N = 1$, the given equations yield

$$\log_3 (\log_4 x) = \log_4 (\log_2 y) = \log_2 (\log_3 z) = 1.$$

Moreover, since $\log_b M = 1$, only if $M = b$, we have

$$\log_4 x = 3, \qquad \log_2 y = 4, \qquad \log_3 z = 2,$$

or equivalently

$$x = 4^3, \qquad y = 2^4, \qquad z = 3^2.$$

Adding these results gives

$$x + y + z = 4^3 + 2^4 + 3^2 = 64 + 16 + 9 = 89.$$

22. (A) Factoring $x^3 - 1 = 0$, which is equivalent to $x^3 = 1$, gives $(x - 1)(x^2 + x + 1) = 0$, so $x - 1 = 0$ or $x^2 + x + 1 = 0$. Since w is imaginary, $w - 1 \neq 0$. Thus $w^2 + w + 1 = 0$ and hence

$$w^2 + 1 = -w, \qquad \text{and} \qquad w + 1 = -w^2.$$

We use these equalities to simplify the given product, getting:

$$(1 - w + w^2)(1 + w - w^2) = (-w - w)(-w^2 - w^2)$$
$$= (-2w)(-2w^2) = 4w^3.$$

But w is a root of $x^3 = 1$, so $w^3 = 1$, and $4w^3 = 4$.

23. (A) Team A may win the series by winning both of two games, or the last of three or four games after winning one of the others. The six possible sequences of wins, each followed by its probability, are $AA(\frac{1}{4})$, $BAA(\frac{1}{8})$, $ABA(\frac{1}{8})$, $BBAA(\frac{1}{16})$, $BABA(\frac{1}{16})$, and $ABBA(\frac{1}{16})$. Since these sequences are mutually exclusive, the sum of their probabilities, 11/16, is the probability of A's winning. Now the *odds* that an event with probability p occurs is defined to be the ratio $p/(1 - p)$; hence the odds favoring Team A are 11 to 5.

We may check the above by computing the complementary odds of 5 to 11 that Team B win the series, which may be accomplished by B winning all of three games or the last of four games after winning two of the first three. The four possible sequences of wins, each followed by its probability, are $BBB(\frac{1}{8})$, $ABBB(\frac{1}{16})$, $BABB(\frac{1}{16})$ and $BBAB(\frac{1}{16})$ with total probability of 5/16 (and hence odds of 5 to 11) favoring Team B to win the series. These odds reversed give the complementary odds of 11 to 5 favoring Team A to win the series as found above.

24. (D) There are $1, 2, 3, \cdots, n$ integers in the 1st, 2nd, 3rd, $\cdots$, nth rows and therefore a total† of

$$1 + 2 + 3 + \cdots + n = \tfrac{1}{2}n(n + 1)$$

integers in the first n rows. Since each row, except the first which has only one, contains two 1's, the number of 1's in the first n rows is $2n - 1$. Therefore the number of integers which are not 1's is

$$\tfrac{1}{2}n(n + 1) - (2n - 1) = \tfrac{1}{2}(n^2 - 3n + 2).$$

The quotient of this number and the number $2n - 1$ of 1's is

$$\frac{\tfrac{1}{2}(n^2 - 3n + 2)}{2n - 1} = \frac{n^2 - 3n + 2}{4n - 2}$$

as stated in choice (D).

OR

† See footnote on p. 114.

We may observe that the number of integers which are not 1's in the 1st, 2nd, 3rd, 4th, $\cdots$, nth rows are $0, 0, 1, 2, \cdots$, $(n - 2)$, respectively, which total

$$0 + 0 + 1 + 2 + \cdots + (n - 2) = \tfrac{1}{2}(n - 2)(n - 1)$$

$$= \tfrac{1}{2}(n^2 - 3n + 2)$$

so that the quotient of this total and the number $2n - 1$ of 1's is $(n^2 - 3n + 2)/(4n - 2)$, as before.

25. (D) Let b and f denote the boy's and his father's age, respectively. The statements in the problem imply: $100f + b - (f - b)$ $= 4289$, that is $99f + 2b = 4289$, or $99f = 4257 + 32 - 2b$, or

(*) $$f - 43 = \frac{32 - 2b}{99}.$$

Suppose, for the moment, that the father's age is 43. Then $2b = 32$, $b = 16$, so the boy is indeed a teenager.

If $f \geq 44$, then $(32 - 2b)/99 \geq 1$, so $b < 0$, which is impossible.

If $f \leq 42$, then $(32 - 2b)/99 \leq -1$, from which we get $32 - 2b \leq -99$, or $2b \geq 131$. Thus $b > 65$, which is hardly teenage. So the only solution satisfying the conditions of the problem is $f = 43$, $b = 16$. Thus $f + b = 59$.

Comment: Solutions by means of congruences (see Solution of Problem 12 of this 1971 Exam., p. 150) are as follows:

(a) Casting out 9's—the equation $99f + 2b = 4289$ reduces to $0 + 2b \equiv 5 \pmod 9$ so that $2b \equiv 5 \pmod 9$, or $b \equiv 7 \pmod 9$. Thus b is one of the numbers $7, 16, 25, 34, \cdots$ of which only 16 is in the teens.

(b) Casting out 11's—The equation $99f + 2b = 4289$ reduces to $0 + 2b \equiv 10 \pmod{11}$, $b \equiv 5 \pmod{11}$, so b is one of the numbers $5, 16, 27, \cdots$, of which only 16 is in the teens.

Substituting 16 for b in $99f + 2b = 4289$ yields $99f = 4257 = 99 \cdot 43$, and $f = 43$.

26. (B) Draw FH parallel to line AGE (see figure). Then $BE = EH$ because $BG = GF$ and a line (GE) parallel to the base (HF) of a triangle (HFB) divides the other two sides proportionally. By the same reasoning applied to triangle AEC with line FH parallel to base AE, we see that $HC = 2EH$, because $FC = 2AF$ is given. Therefore $EC = EH + HC = 3EH = 3BE$, and E divides side BC in the ratio 1:3.

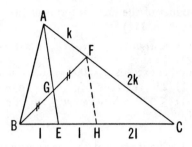

27. (E) Let r, w, and b denote the numbers of red, white, and blue chips, respectively. We are given that $\frac{1}{2}w \leq b \leq \frac{1}{3}r$, and that $55 \leq w + b$. Now since $w \leq 2b$, $55 \leq 2b + b = 3b$. Hence $b \geq 55/3 = 18\frac{1}{3}$, and since b is an integer, this implies that $b \geq 19$. But by hypothesis, $r \geq 3b$. Hence $r \geq 57$.

28. (C) Let b and h denote the lengths of the base and altitude, respectively, of the given triangle. Then the largest of the ten parts into which the triangle is divided is a trapezoid with altitude of length $.1h$ and parallel bases of lengths b and $.9b$. The given area of this largest part is

$$\tfrac{1}{2}(.1h)(b + .9b) = .19(\tfrac{1}{2}bh) = 38$$

from which the area of the given triangle is $\frac{1}{2}bh = 200$.

OR

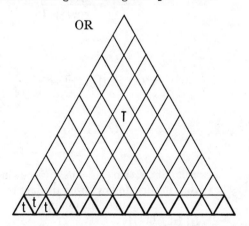

Through the points where the parallel lines intersect each side of the given triangle T, draw lines parallel to the other sides (see figure). Thus the trapezoidal strip of area 38 is subdivided into 19 congruent triangles t, so the area of each is 2. Since the sides of T are 10 times the corresponding sides of t,

Area $T = 100$ Area $t = 200$.

29. (E) The product of the first n terms of the given progression (see footnote on p. 114) is

$$10^{1/11}\cdot 10^{2/11}\cdot 10^{3/11}\cdot \ \cdots\ \cdot 10^{n/11} = 10^{(1+2+3+\cdots+n)/11} = 10^{n(n+1)/22}.$$

This number exceeds $100{,}000 = 10^5$ if and only if the exponent $n(n+1)/22$ exceeds 5, i.e. $n(n+1) > 110$. Since $n(n+1) \le 110$ whenever $n \le 10$, the required least integer n is 11.

30. (D) Let g be the inverse of the transformation f_1; then $f_1[g] = g[f_1]$ is the identity transformation, and $g[f_{n+1}(x)] = g[f_1(f_n(x))] = f_n(x)$. Repeated application of g k times yields

$$g(g(g(\cdots(f_{n+1}(x))\cdots) = g^k f_{n+1}(x) = f_{n+1-k}(x),$$

and applying g five times to the given identity $f_{35}(x) = f_5(x)$ yields

$$g^5 f_{35}(x) = f_{30}(x) = g^5 f_5(x) = x,$$

so that f_{30} is the identity map. It follows that

$$f_{28}(x) = g^2\{f_{30}(x)\} = g^2(x) = g[g(x)].$$

Since

$$f_1[g(x)] = \frac{2g(x) - 1}{g(x) + 1} = x, \qquad g(x) = \frac{x + 1}{2 - x},$$

and

$$g^2(x) = g[g(x)] = \frac{g + 1}{2 - g} = \frac{\dfrac{x + 1}{2 - x} + 1}{2 - \dfrac{x + 1}{2 - x}}$$

$$= \frac{x + 1 + 2 - x}{4 - 2x - x - 1} = \frac{3}{3 - 3x} = \frac{1}{1 - x}.$$

Comment: Show that $f_{35} = f_5$ by showing that f_{30} is the identity; this, in turn, follows from the fact that f_6 is the identity, and f_6 can be computed, for example, from $f_6(x) = f_1\{f_1(f_1(x))\} = f_2\{f_2(f_2(x))\}$.

Part 4

31. (A) Radius OB bisects chord AC at right angles in point G. Since CD also makes a right angle with AC, $CD \parallel BO$. Angles ADB

and BAG are equal because they are measured by half the equal arcs AB and BC. Hence right triangles ABD and BGA are similar with $BG/AB = AB/AD$, so $BG = \frac{1}{4}$ and $OG = OB - BG = 2 - \frac{1}{4} = \frac{7}{4}$. Since $CD \parallel GO$, $CD/GO = AD/AO = 2$, so $CD = 2 \cdot \frac{7}{4} = \frac{7}{2}$.

OR

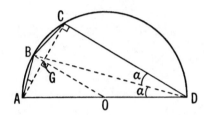

Denote $\angle ADB$ by α; then $\angle ADC = 2\alpha$. Using right triangles ABD and ACD, we have

$$\sin \alpha = \frac{1}{4}, \quad \cos 2\alpha = \cos^2 \alpha - \sin^2 \alpha = 1 - 2\sin^2 \alpha = \frac{CD}{4}.$$

$$\therefore CD = 4(1 - 2 \cdot \tfrac{1}{16}) = 4 - \tfrac{1}{2} = \tfrac{7}{2}.$$

32. (A) In factored form

$$1 - x^{32} = (1 - x)(1 + x)(1 + x^2)(1 + x^4)(1 + x^8)(1 + x^{16}).$$

When we set $x = 2^{-1/32}$, this equality becomes $1 - 2^{-1} = (1 - 2^{-1/32})s$ from which $s = \frac{1}{2}(1 - 2^{-1/32})^{-1}$.

Note: In evaluating the product

$$(1 + x)(1 + x^2)(1 + x^4)(1 + x^8)(1 + x^{16})$$

we find the sum of all terms obtainable by taking one number from each parenthesis and multiplying them together:

$$1^5 + 1^4 x + 1^4 x^2 + 1^3 x \cdot x^2 + \cdots + x^{1+2+4+\cdots+16}.$$

Our sum consists of terms x^k for all those k which can be written as sums of distinct powers of 2 from the zero-th to the fourth power. By the properties of the binary system, every integer k with $0 \le k \le 31$ has a unique such representation. Accordingly, the product above is equal to

$$1 + x + x^2 + x^3 + \cdots + x^{31},$$

and this geometric series has the value $(1 - x^{32})/(1 - x)$, whence $1 - x^{32} = (1 - x) \cdot$ given product. This motivates the above solution to some extent.

33. (B) The product P of the n quantities in G.P. which we denote by $a, ar, ar^2, \cdots, ar^{n-1}$ is given by†

$$P = a \cdot ar \cdot ar^2 \cdot \ \cdots \ \cdot ar^{n-1} = a^n r^{1+2+\cdots+n-1} = a^n r^{n(n-1)/2}.$$

The sum S of the n quantities is

$$S = a + ar + ar^2 + \cdots + ar^{n-1} = a(1 - r^n)/(1 - r).$$

The sum S' of the reciprocals of the n quantities is

$$S' = \frac{1}{a} + \frac{1}{ar} + \cdots + \frac{1}{ar^{n-1}} = \frac{1}{a}\frac{1 - r^{-n}}{1 - r^{-1}} = \frac{1}{a}\cdot\frac{1}{r^{n-1}}\cdot\frac{1 - r^n}{1 - r}.$$

The quotient of S and S' is the product of S and the reciprocal of S':

$$\frac{S}{S'} = \frac{a(1 - r^n)}{1 - r}\cdot\frac{ar^{n-1}(1 - r)}{1 - r^n} = a^2 r^{n-1}.$$

If we raise this quantity to the power $n/2$, we obtain

$$\left(\frac{S}{S'}\right)^{n/2} = (a^2 r^{n-1})^{n/2} = a^n r^{n(n-1)/2} = P.$$

34. (B) In a correctly running clock the minute hand moves $6°$ per minute, the hour hand moves $\frac{1}{2}°$ per minute. Suppose both hands coincide. After x minutes, the hands have travelled $6x$ and $x/2$ degrees, respectively, and will coincide again when $6x - 360° = x/2$, that is, in $x = 720/11 = 65\frac{5}{11}$ minutes. Thus the ratio of the time indicated by the slow clock to the true time is

$$\frac{720/11}{69} = \frac{720}{11\cdot69} = \frac{240}{11\cdot23} = \frac{240}{253}.$$

When the slow clock indicates 8 hours = 480 minutes, the true time t is obtained from

$$\frac{480}{t} = \frac{240}{253}, \qquad t = \frac{480}{240}\cdot253 = 2\cdot253 = 506 = 480 + 26.$$

Thus 26 minutes are lost in the false eight hour recording. At time and a half of $4 per hour, i.e. at $6 per hour or 10 cents per minute, the extra pay should be $2.60 for 26 minutes.

† See footnote on p. 114.

OR

We note that 12 hours by the slow clock = 11·69 minutes = 12 hours + 39 minutes. Therefore 8 hours by the slow clock = 8 hours + 26 minutes. Thus 26 minutes is lost by the slow clock and the extra pay should be $2.60 as before.

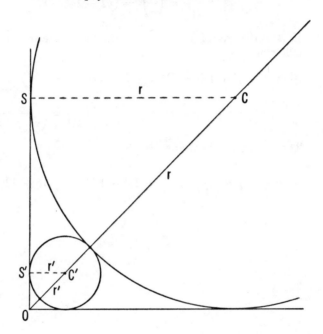

35. (C) Let O denote the vertex of the given right angle (see figure), C and C' the centers, r and r' $(r > r')$ the radii of the larger and smaller, respectively, of any two consecutive circles in the infinite sequence. Let CS and $C'S'$ be radii perpendicular to one side of the right angle. Then OCS and $OC'S'$ are isosceles right triangles, and $OC = \sqrt{2}r$, $OC' = \sqrt{2}r'$. Moreover, the distance from O to the point of tangency of the two circles with each other, is

$$OC - r = \sqrt{2}r - r = (\sqrt{2} - 1)r \quad \text{in terms of } r, \text{ and}$$

$$OC' + r' = \sqrt{2}r' + r' = (\sqrt{2} + 1)r' \quad \text{in terms of } r'.$$

Equating these equal expressions we obtain

$$(\sqrt{2} + 1)r' = (\sqrt{2} - 1)r \quad \text{so that} \quad \frac{r'}{r} = \frac{\sqrt{2} - 1}{\sqrt{2} + 1} = (\sqrt{2} - 1)^2.$$

We note in passing that $(\sqrt{2} - 1)$ and $(\sqrt{2} + 1)$ are reciprocals, i.e. $(\sqrt{2} - 1)(\sqrt{2} + 1) = 1$.

Now the ratio of the areas of two consecutive circles is the square of the ratio of their radii:

$$\frac{\pi r'^2}{\pi r^2} = \left(\frac{r'}{r}\right)^2 = (\sqrt{2} - 1)^4.$$

If A is the area of the first circle, the sum of the areas of all the following is

$$A[(\sqrt{2} - 1)^4 + (\sqrt{2} - 1)^8 + (\sqrt{2} - 1)^{12} + \cdots]$$

$$= A \frac{(\sqrt{2} - 1)^4}{1 - (\sqrt{2} - 1)^4} = \frac{A}{(\sqrt{2} + 1)^4 - 1}.$$

The required ratio of areas is

$$A : \frac{A}{(\sqrt{2} + 1)^4 - 1} = [(\sqrt{2} + 1)^4 - 1] : 1 = (16 + 12\sqrt{2}) : 1.$$

1972 Solutions

Part 1

1. (D) An extension of the Pythagorean Theorem states that the angle opposite the longest side of a triangle is acute, right, or obtuse according as the square on that side is less than, equal to, or greater than the sum of the squares on the other two sides. For the given triangles a tabulation follows:

	Sides	(Longest Side)^2	$\begin{matrix} < \\ = \\ > \end{matrix}$	Sum of Squares	Angle Opposite
I	3, 4, 5	25	=	9 + 16	Right
II	4, $7\frac{1}{2}$, $8\frac{1}{2}$	$72\frac{1}{4}$	=	16 + $56\frac{1}{4}$	Right
III	7, 24, 25	625	=	49 + 576	Right
IV	$3\frac{1}{2}$, $4\frac{1}{2}$, $5\frac{1}{2}$	$30\frac{1}{4}$	<	$16\frac{1}{4}$ + $20\frac{1}{4}$	Acute

We see that I, II, and III are the only right triangles.

2. (B) Let C denote the present cost so that $.92C$ is the cost at 8% less. Since selling price is cost plus profit, selling price = cost + x%·cost. Equating the selling price with cost C at x% profit and that with cost $.92C$ at $(x + 10)$% profit yields

$$C(1 + .01x) = .92C[1 + .01(x + 10)],$$
$$.08(.01x) = .012, \quad x = 15.$$

3. (B) Straightforward calculation yields the value of

$$x^2 - x = \tfrac{1}{4}(1 - i\sqrt{3})^2 - \tfrac{1}{2}(1 - i\sqrt{3})$$
$$= \tfrac{1}{4}(-2 - 2i\sqrt{3}) - \tfrac{1}{2}(1 - i\sqrt{3}) = -1.$$

The required reciprocal of $x^2 - x$ is -1 or choice (B).

                    OR

We write $x^2 - x = x(x - 1)$ and observe that

$$x - 1 = -\left(\frac{1 + i\sqrt{3}}{2}\right) = -\bar{x},$$

where $\bar{x}$ is the conjugate of x. Thus

$$x(x-1) = -x\bar{x} = -\frac{1}{4}(1+3) = -1, \quad \text{so} \quad \frac{1}{x^2-x} = -1.$$

4. (D) Each set X satisfying the given relation must contain the subset $\{1, 2\}$ and also be a subset of $\{1, 2, 3, 4, 5\}$. These sets X are $\{1, 2\}$, $\{1, 2, 3\}$, $\{1, 2, 4\}$; $\{1, 2, 5\}$, $\{1, 2, 3, 4\}$, $\{1, 2, 3, 5\}$, $\{1, 2, 4, 5\}$ and $\{1, 2, 3, 4, 5\}$. The number of sets X is 8 or choice (D).

OR

The sets X are each the union of the set $\{1, 2\}$ with a subset of $\{3, 4, 5\}$. The number of these subsets is 8; they include 6 proper subsets, the empty set, and the entire set $\{3, 4, 5\}$.

5. (A) First notice that $2^{1/2} > 8^{1/8}$ because $(2^{1/2})^8 = 2^4 = 16$ exceeds $(8^{1/8})^8 = 8$. Also $2^{1/2} > 9^{1/9}$ because $(2^{1/2})^{18} = 2^9 = 512$ exceeds $(9^{1/9})^{18} = 9^2 = 81$. Moreover, $3^{1/3} > 2^{1/2}$ because $(3^{1/3})^6 = 3^2 = 9$ exceeds $(2^{1/2})^6 = 2^3 = 8$. Now since $3^{1/3} > 2^{1/2}$ and $2^{1/2}$ exceeds both $8^{1/8}$ and $9^{1/9}$, therefore $3^{1/3}$ and $2^{1/2}$ are the greatest and the next to the greatest of the four given numbers, in that order.

Remark: Our method of comparing $a^{1/p}$ and $b^{1/q}$ has been to raise each to the power $k = $ least common multiple of p, q, and to compare the resulting numbers $a^{k/p}$, $b^{k/q}$ whose exponents are integers. Comparison of $8^{1/8}$ and $9^{1/9}$ (not needed in this problem) by this method would lead to comparing $(8^{1/8})^{72} = 8^9$ and $(9^{1/9})^{72} = 9^8$ and can be accomplished as follows.

We have $2^5 = 32 > 27 = 3^3$. Hence $2^{25} > 3^{15}$. Since $2^2 = 4 > 3$, this implies that $2^{27} > 3^{16}$, or in other words that $8^9 > 9^8$.

Note: The given numbers are of the form $n^{1/n}$, and the problem raises the question: Where does the function $f(x) = x^{1/x}$ increase, where does it decrease as $x > 0$ increases? This question can be answered by methods of calculus. First we note that $\log f(x)$ increases or decreases according as $f(x)$ increases and decreases; secondly, that functions increase or decrease according as their derivatives are positive or negative. Thus

$$\log_e f(x) = \frac{1}{x}\log_e x,$$

$$\frac{d}{dx}\left(\log_e f(x)\right) = \frac{1}{x^2}\left(1 - \log_e x\right)\begin{cases}>0 \text{ for } \log_e x < 1, \quad x < e \\ <0 \text{ for } \log_e x > 1, \quad x > e,\end{cases}$$

where e, the base of natural logarithms, lies between 2 and 3. Thus we see that for $m > n \ge 3$, $m^{1/m} < n^{1/n}$; but $2^{1/2}$ and $3^{1/3}$ have to be compared because the maximum of $x^{1/x}$ occurs at $2 < x = e < 3$.

6. (C) Set $y = 3^x$ so that the given equation is equivalent to

$$y^2 - 10y + 9 = 0 \quad \text{or} \quad (y - 9)(y - 1) = 0.$$

Hence $y = 3^x = 9$ or 1, so that $x = 2$ or 0. Therefore $x^2 + 1 = 2^2 + 1 = 5$ or $x^2 + 1 = 0^2 + 1 = 1$ as stated in choice (C).

7. (E) The required ratio is

$$\frac{x}{yz}\Big/\frac{y}{zx} = \frac{x}{yz}\cdot\frac{zx}{y} = \frac{x^2}{y^2}.$$

The given information includes the fact that $yz/zx(=y/x) = 1/2$, so $x/y = 2/1$ and $x^2/y^2 = 4/1$ is the required ratio. The other piece of given information, $zx/xy = 2/3$, is unnecessary to solve the problem.

8. (D) The difference $x - \log y$ is either non-negative or negative so that the given equation requires that either

$$x - \log y = x + \log y, \quad 2\log y = 0, \quad y = 1$$

or

$$-(x - \log y) = x + \log y, \quad 2x = 0, \quad x = 0.$$

We can write $x(y - 1) = 0$ to say that either $y = 1$, or $x = 0$, or both.

Comment: We note that the solution set of the given equation is $\{(x, y) \mid x = 0 \text{ and } y \ge 1, \text{ or } y = 1 \text{ and } x \ge 0\}$, which is a subset of the set of all (x, y) for which $x(y - 1) = 0$. We also note that without even solving the problem one can eliminate choices (A), (B), and (C), since they all imply (D), and by assumption only one choice is correct.

9. (A) Let S and E denote the number of sheets of paper and of envelopes, respectively, in each box. Then addition of the resulting equations

$$S - E = 50 \quad \text{and} \quad E - S/3 = 50$$

gives $\frac{2}{3}S = 100$, $S = 150$ sheets of paper in each box.

10. (D) The value of $(x - 2)$ is either positive or negative, and then the given inequality is equivalent respectively to

$$1 \leq x - 2 \leq 7, \quad 3 \leq x \leq 9$$

or

$$1 \leq 2 - x \leq 7, \quad -1 \leq -x \leq 5, \quad -5 \leq x \leq 1.$$

These two alternatives are equivalent to the inequalities stated in choice (D).

Part 2

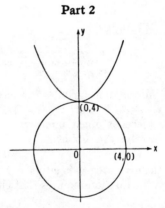

11. (A) From the graph of the first equation (a circle with center at the origin and radius 4; see figure), and that of the second equation, a parabola, symmetric with respect to the y-axis, concave upward and having its vertex at $(0, 4)$, it is apparent that the only common point of both graphs is $x = 0$, $y = 4$; so $y = 4$ is the only admissible value of y.

OR

We solve the equations algebraically, substituting $x^2 = 3y - 12$ from the second equation into the first;

$$3y - 12 + y^2 - 16 = 0,$$

or

$$y^2 + 3y - 28 = (y - 4)(y + 7) = 0,$$

$y = 4$ or $y = -7$. Clearly, $y = 4$, $x = 0$ satisfies both given equations while $y = -7$ satisfies neither no matter what real x we take.

12. (B) Let an edge of the cube be f feet and hence $12f$ inches long. Equating the number of cubic feet in the volume to the number of square inches in all 6 faces, we get $f^3 = 6(12f)^2$ so that $f = 6 \cdot (12)^2 = 864$.

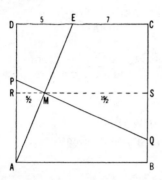

13. (C) Let the line through M parallel to side AB of the square intersect sides AD and BC in points R and S, respectively; see figure. Since M is the midpoint of AE, $RM = \frac{1}{2}DE = \frac{5}{2}$ inches, and hence $MS = 12 - \frac{5}{2} = \frac{19}{2}$ inches. Since PMR and QMS are similar right triangles, the required ratio

$$PM:MQ = RM:MS = 5:19$$

because corresponding sides of similar triangles are proportional.

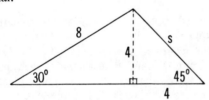

14. (B) Let s denote the length of the required side (see figure). Then the altitude to the longest side, opposite the 30° angle, has length $8/2 = 4$ and is one leg of an isosceles right triangle with hypotenuse s, which therefore has length $4\sqrt{2}$.

OR

By the Law of Sines which states that the sides of any triangle are proportional to the sines of the angles opposite them, we have

$$\frac{s}{\sin 30°} = \frac{8}{\sin 45°} \quad \text{or} \quad s = \frac{8 \sin 30°}{\sin 45°} = \frac{8(1/2)}{\sqrt{2}/2} = 4\sqrt{2}$$

as before.

15. (C) Let x denote the number of bricks in the wall; then $x/9$ and $x/10$ bricks per hour would be laid by each bricklayer if he worked alone. Working together they lay 10 fewer or $(x/9)$ ⊥ $(x/10) - 10$ bricks per hour. Now since x bricks are laı 5 hours, we have $5[x/9 + (x/10) - 10] = x$, so there $x = 900$ bricks in the wall.

16. (B) Let the positive numbers be denoted by x and y with the first three $3, x, y$ and the last three $x, y, 9$. Then

$$x/3 = y/x \quad \text{and} \quad y - x = 9 - y,$$

because the first three are in geometric and the last three in arithmetic progression. Eliminating y from these two equations, we get

$$2x^2 - 3x - 27 = 0 \quad \text{or} \quad (2x - 9)(x + 3) = 0,$$

$$x = 9/2 \quad \text{or} \quad -3.$$

Since x is required to be positive, we use $x = 9/2$ to find $y = 27/4$, and hence the required sum is $x + y = 45/4 = 11\frac{1}{4}$.

17. (E) To select a cutting point at random means that the probability of the cutting point falling within a given interval is proportional to the length of that interval. Let AB represent the string (see figure) and let P be the point such that $AP/PB = 1/x$. Now if the cut lies on AP, the longer piece is at least x times as large as the shorter. The probability of the cut being on AP is $1/(1 + x)$. Now the random cut is equally likely to lie within the same distance from the other end B of the string so that the probability is $2/(1 + x)$.

Remark: Choices (A), (B), and (C) are immediately eliminated by the fact that when $x = 1$, the required probability is clearly 1.

18. (A) We may extend sides AD and BC of the trapezoid to meet at V; see figure. Then AC and BD are medians from vertices A and B of $\triangle ABV$ meeting at point E, which divides the length of each in the ratio $2:1$. This means that

$$EC = \tfrac{1}{3}AC = \tfrac{11}{3} = 3\tfrac{2}{3}.$$

OR

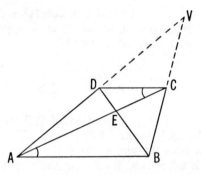

Since base AB is parallel to base CD, alternate interior angles BAC and DCA are equal, so that $\triangle ABE$ is similar to $\triangle CDE$. We are given that base $AB = 2DC$ so that the corresponding sides AE and EC are such that $AE = 2EC$. Moreover diagonal $AC = AE + EC = 2EC + EC = 3EC = 11$. Hence $EC = 11/3 = 3\frac{2}{3}$.

19. (D) The kth term $(1 + 2 + 2^2 + \cdots + 2^{k-1})$ of the given sequence is a geometric series whose value is $2^k - 1$. The sum of the first n terms is therefore.

$$(2^1 - 1) + (2^2 - 1) + (2^3 - 1) + \cdots + (2^n - 1)$$
$$= (2^1 + 2^2 + 2^3 + \cdots + 2^n) - \underbrace{(1 + 1 + 1 + \cdots + 1)}_{n \text{ terms}}$$

$$= (2^{n+1} - 2) - n = 2^{n+1} - n - 2 \qquad \text{or choice (D)}.$$

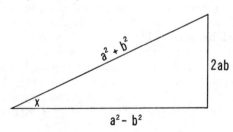

20. (E) The acute angle x may be taken as opposite the leg of length $2ab$ in a right triangle with other leg of length $(a^2 - b^2)$. Then the square of the hypotenuse h is, by the Pythagorean Theorem,

$$h^2 = (2ab)^2 + (a^2 - b^2)^2 = a^4 + 2a^2b^2 + b^4 = (a^2 + b^2)^2.$$

We now see from the figure and the definition of sine that $\sin x = 2ab/(a^2 + b^2)$.

Remark: Since $\tan x = \sin x/\cos x = 2ab/(a^2 - b^2)$, we see that $\tan x$ and $\sin x$ both approach 0 as b approaches 0. This eliminates choices (A), (C) and (D).

Part 3

21. (C) Let P and Q be the intersections of AD with BF and EC, respectively, and denote $\angle FPQ$ by $\angle P$ and $\angle EQP$ by $\angle Q$. Then since the sum of the angles of quadrilateral $EFPQ$ is 360° (see figure) and the sum of the angles of $\triangle DPB$ and $\triangle AQC$ are each 180°, we have the 3 equations

$$\angle F + \angle P + \angle Q + \angle E = 360°$$
$$\angle B + (180° - \angle P) + \angle D = 180°$$
$$\angle C + (180° - \angle Q) + \angle A = 180°.$$

Addition of these equations gives, after subtracting 360° from each side, the required sum

$$\angle A + \angle B + \angle C + \angle D + \angle E + \angle F = 360° = 90n°$$

so that n finally is equal to 4 or choice (C).

OR

Starting with a rigid stick in the direction AD, we can pivot it around A in the counterclockwise direction until it lies in the direction of AC. We then pivot it around C to make it lie along CE, then around E into the direction EF, then around F into FB, then around B into BD, and finally around D until the stick again lies along AD. It is clear that the stick has rotated by $k \cdot 180°$, where k is an integer. In this particular example, the stick has made one full turn, so it has rotated 360° which is precisely the sum of the measures of all the angles of the polygon; for at each corner, the stick turned by the measure of an angle, and it has turned all the corners. Hence the sum of the angles is $360° = 4 \cdot 90°$, so $n = 4$.

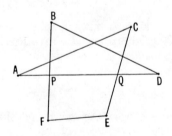

22. (E) For convenience, abbreviate the given root $a + bi$ by α and its conjugate $a - bi$ by $\bar{\alpha}$. Satisfaction of the given equation by each of these roots requires that

$$\alpha^3 + q\alpha + r = 0 \qquad \text{and} \qquad \bar{\alpha}^3 + q\bar{\alpha} + r = 0.$$

Subtracting the second from the first of these equations yields

$$\alpha^3 - \bar{\alpha}^3 + q(\alpha - \bar{\alpha}) = 0$$

from which

$$-q = \frac{\alpha^3 - \bar{\alpha}^3}{\alpha - \bar{\alpha}} = \alpha^2 + \alpha\bar{\alpha} + \bar{\alpha}^2 = (\alpha^2 + \bar{\alpha}^2) + \alpha\bar{\alpha}$$

$$= (2a^2 - 2b^2) + (a^2 + b^2) = 3a^2 - b^2.$$

Hence $q = b^2 - 3a^2$ as stated in choice (E).

OR

We may note that since the coefficient of x^2 in the original equation is zero, so also is the sum of the three roots. The third root of the equation is therefore $-2a$. Now the coefficient of x is the sum of the products of the roots taken two at a time:

$$q = (a + bi)(a - bi) + (-2a)(a + bi) + (-2a)(a - bi)$$

$$= a^2 + b^2 - 2a^2 - 2abi - 2a^2 + 2abi = b^2 - 3a^2,$$

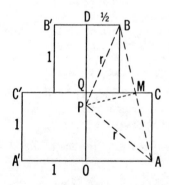

23. (D) Let O denote the center of the base of the figure, P the center of the desired circle and r its radius. We assume that the circle passes through the points labelled A and B, and that the center P is on the axis of symmetry OD of the figure; this assumption is justified in the comment below. Thus $r = PA = PB$. In right triangle PDB, $PB^2 = (2 - OP)^2 + (\frac{1}{2})^2$, and in right triangle POA, $PA^2 = OP^2 + 1$. Equating these expressions for r^2 yields

$$4 - 4OP + OP^2 + \tfrac{1}{4} = OP^2 + 1,$$

so

$$4OP = \tfrac{13}{4} \quad \text{and} \quad OP = \tfrac{13}{16}$$

whence

$$r^2 = 1 + OP^2 = \frac{425}{16^2} \quad \text{and} \quad r = \frac{5\sqrt{17}}{16}.$$

OR

Draw chord AB; its midpoint M lies on $C'C$, and its perpendicular bisector MP is the hypotenuse of right $\triangle PQM$, which is similar to $\triangle MCA$ with

$$\frac{PQ}{QM} = \frac{MC}{CA} \quad \text{or} \quad \frac{PQ}{3/4} = \frac{1/4}{1}.$$

Hence

$$PQ = \frac{3}{16}, \quad OP = 1 - \frac{3}{16} = \frac{13}{16}, \quad \text{and} \quad r = \frac{5\sqrt{17}}{16}$$

as before.

OR

Place $A'A$ on the x-axis and O at the origin of a rectangular coordinate plane and determine the ordinate $k(=OP)$ of P and the radius r from the equation $x^2 + (y - k)^2 = r^2$ of the circle, using the fact that the coordinates of A: $(1, 0)$ and B: $(\tfrac{1}{2}, 2)$ satisfy it.

Comment: We suggest that the reader verify the following two facts:

I The smallest circle K having a given polygon in its interior passes through some of the vertices of the polygon; moreover, not all the vertices lying on K lie on a minor arc of K.

II If the given polygon has a line of symmetry, the center P of K lies on it.

To prove I, show that, if no vertices were on K, or if all vertices on K were on a minor arc of K, a circle smaller than K would contain the figure. To prove II, use I.

It follows from II that in our problem, P lies on OD; and it follows from I that P lies on the segment s connecting the midpoint of OQ to Q. For, if P were any higher, the circle would pass either through no vertices or through A and A' only, so A and A' would lie on a minor arc contradicting I.

If P were any lower, K would pass through either no vertices or through B and B' only, and these would lie on a minor arc, again contradicting I.

For any point S on segment s, $SA > SC$ and $SB > SC$. Hence A and B are on the circle, as we surmised.

24. (B) All three walking rates and corresponding times yield the same distance. If R, T denote the first rate and corresponding time, we have the following expressions for the distance:

$$RT = (R + \tfrac{1}{2})\tfrac{4}{5}T = (R - \tfrac{1}{2})(T + \tfrac{5}{2}).$$

The first equality is equivalent to

$$R = \tfrac{4}{5}(R + \tfrac{1}{2}), \quad \text{so} \quad R = 2, \quad \text{and} \quad RT = 2T.$$

Using the last expression for the distance, we get

$$RT = \left(R - \frac{1}{2}\right)\left(T + \frac{5}{2}\right) = \left(2 - \frac{1}{2}\right)\left(\frac{RT}{2} + \frac{5}{2}\right) = \frac{3RT}{4} + \frac{15}{4},$$

whence

$$\frac{RT}{4} = \frac{15}{4}, \quad RT = 15 = \text{distance in miles}.$$

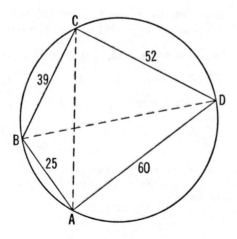

25. (C) The circle circumscribed about quadrilateral $ABCD$ (see figure) is the circumcircle of each of the triangles BAD, BCD, ABC, ADC. The diameter $2R$ of the circumcircle of any triangle is equal to the length of any side divided by the sine

of the angle opposite that side.† Triangles BAD and BCD have side BD in common, and angles A and C opposite BD are opposite angles of the inscribed quadrilateral; hence $\angle A + \angle C = 180°$ and $\cos C = -\cos A$. By the law of cosines,

(1)
$$BD^2 = AB^2 + AD^2 - 2AB \cdot AD \cos A$$
$$= CB^2 + CD^2 - 2CB \cdot CD \cos C,$$

so

$$AB^2 + AD^2 - 2AB \cdot AD \cos A = CB^2 + CD^2 + 2CB \cdot CD \cos A,$$

whence

(2)
$$\cos A = \frac{CB^2 + CD^2 - AB^2 - AD^2}{2(CB \cdot CD + AB \cdot AD)}.$$

Once $\cos A$ is determined, BD can be found from (1) and so can $2R = BD/\sin A = BD/\sqrt{1 - \cos^2 A}$. This general set-up works equally well for triangles ABC and ADC with common side AC and leads to

(3)
$$\cos D = \frac{DA^2 + DC^2 - BA^2 - BC^2}{2[DA \cdot DC + BA \cdot BC]}.$$

The given lengths in our problem are such that an alert reader may discover a shortcut due to these relationships:

$$BC = 39 = 3 \cdot 13, \qquad CD = 52 = 4 \cdot 13,$$

$$AB = 25 = 5 \cdot 5, \qquad DA = 60 = 12 \cdot 5;$$

$$BC^2 + CD^2 = 13^2[3^2 + 4^2] = 13^2 \cdot 5^2,$$

$$AB^2 + DA^2 = 5^2[5^2 + 12^2] = 5^2 \cdot 13^2$$

---

† Let J be the other end of the diameter through C. $\angle JBC$ is right, and $\angle J = \angle A$. Hence $\sin A = \sin J = a/2R$, so $2R = a/\sin A$. Relations $2R = b/\sin B = c/\sin C$ are obtained analogously.

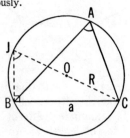

so that
$$BC^2 + CD^2 = AB^2 + DA^2$$

which means that triangles BAD and BCD are right with common hypotenuse BD of length $BD = \sqrt{5^2 \cdot 13^2} = 65$, and $BD = 2R$ is the diameter of the desired circle.

Had we inserted the given data into formula (2), we should have found that $\cos A = 0$ and concluded that $\angle A = \angle C = 90°$, discovering that $BD = 2R$. On the other hand, using formula (3), we obtain

$$\cos D = \frac{60^2 + 52^2 - 25^2 - 39^2}{2[60 \cdot 52 + 25 \cdot 39]} = \frac{5^2(12^2 - 5^2) + 13^2(4^2 - 3^2)}{2 \cdot 5 \cdot 13[63]}$$

$$= \frac{5^2 \cdot 7 \cdot 17 + 13^2 \cdot 7}{2 \cdot 3^2 \cdot 5 \cdot 7 \cdot 13} = \frac{33}{65}$$

whence $AC^2 = 52^2 + 60^2 - 2 \cdot 52 \cdot 60 \cos D = 56^2$ and

$$2R = \frac{AC}{\sin D} = \frac{56}{\sqrt{1 - (33/65)^2}} = \frac{56}{56/65} = 65 \quad \text{as before.}$$

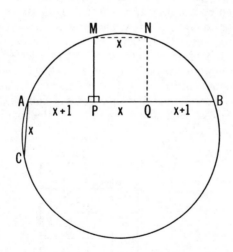

26. (E) Locate point N on arc AMB between M and B so that arcs CA and MN are equal (see figure). Then arcs AM and BN are equal, and hence segment NQ drawn $\perp$ to AB at Q is parallel and equal in length to MP. Now MN and PQ are opposite sides of rectangle $MNPQ$, and each has measure x. Hence segment PB has measure $PQ + QB$ $= x + (x + 1) = 2x + 1$, because $QB = AP = x + 1$.

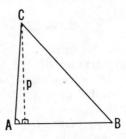

27. (D) The area of any triangle is equal to half the product of any two sides and the sine of their included angle, see figure, where p is the altitude from C, so $p = AC \sin A$, and

$$\text{Area } \triangle ABC = \tfrac{1}{2}AB \cdot p = \tfrac{1}{2}AB \cdot AC \sin A.$$

From the given data, $\sqrt{AB \cdot AC} = 12$ and Area $\triangle ABC = 64$, so $AB \cdot AC \sin A = 128$, $AB \cdot AC = 144$,

$$\sin A = \tfrac{128}{144} = \tfrac{8}{9}.$$

28. (E) None of the $28 = 8 + 8 + 6 + 6$ border squares is entirely covered by the disc. In the 6×6 checkerboard formed by the interior squares, the four corner squares are not entirely covered, because the distance from each corner of this reduced checkerboard to the center of the disc is $\sqrt{3^2 + 3^2} \cdot D/8 = 3\sqrt{2}D/8$ while the radius of the disc is $D/2$, and $1/2 < 3\sqrt{2}/8$. The remaining $36 - 4 = 32$ interior squares are entirely covered by the disc, since they lie in a circle of radius $\sqrt{2^2 + 3^2}D/8 = \sqrt{13}D/8 < D/2$ about the center of the board.

OR

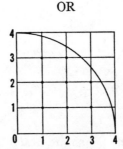

We take the center of the disc as origin O of a rectangular coordinate system with axes along sides of checkerboard squares and with $D/8$ as unit of length. We shall count the number of covered squares in the first quadrant only, see figure, and multiply our result by 4. If we associate with each covered square its upper right-hand vertex, we need only count these vertices. They consist of those points (x, y) with integer coordinates (called lattice points) for which $x > 0$, $y > 0$, and $x^2 + y^2 \le 4^2$. These are the eight points $(1, 1)$, $(1, 2)$, $(1, 3)$, $(2, 1)$, $(2, 2)$, $(2, 3)$, $(3, 1)$, $(3, 2)$. Thus there are $4 \cdot 8 = 32$ covered squares in all four quadrants.

29. (C) Direct calculation yields the desired result

$$f\left(\frac{3x + x^3}{1 + 3x^2}\right) = \log \frac{1 + \dfrac{3x + x^3}{1 + 3x^2}}{1 - \dfrac{3x + x^3}{1 + 3x^2}} = \log \frac{1 + 3x^2 + 3x + x^3}{1 + 3x^2 - 3x - x^3}$$

$$= \log \frac{(1 + x)^3}{(1 - x)^3} = 3 \log \frac{1 + x}{1 - x} = 3f(x).$$

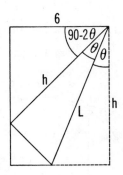

30. (A) In the figure, where h denotes the length of the sheet,

$$6/h = \cos (90° - 2\theta) = \sin 2\theta = 2 \sin \theta \cos \theta$$

from which $h = 3/(\sin \theta \cos \theta)$. Also $L/h = \sec \theta$ and therefore

$$L = h \sec \theta = 3 \sec \theta/(\sin \theta \cos \theta) = 3 \sec^2 \theta \csc \theta$$

or choice (A).

Remark: Choices (B), (C), and (D) can all be eliminated at once from the fact that when the paper is square, we have $\theta = 45°$ and $L = 6\sqrt{2}$.

Part 4

31. (C) We make use of the following fact: If N_1, N_2 are integers whose remainders, upon division by D, are R_1 and R_2, then the products N_1N_2 and R_1R_2 have the same remainder upon division by D. In symbols: If $N_1 = Q_1D + R_1$ and $N_2 = Q_2D + R_2$, then

$$N_1N_2 = (Q_1D + R_1)(Q_2D + R_2)$$
$$= (Q_1Q_2D + Q_1R_2 + Q_2R_1)D + R_1R_2,$$

and the last expression clearly has the same remainder as R_1R_2.

Among the first few powers of two, we find that 2^6 has the convenient remainder 12 [or -1] upon division by 13, so $2^{12} = 2^6 \cdot 2^6$ has the same remainder as $12 \cdot 12$ [or $(-1)^2$], namely 1. We now write $2^{1000} = (2^{12})^{83} \cdot 2^4$, and conclude that the remainder upon division by 13 is $(1)^{83} \cdot 3 = 3$, since $2^4 = 16 = 1 \cdot 13 + 3$.

Using the notation of congruences, we have

$$2^6 = 64 \equiv -1 \ (\text{mod } 13),$$

$$2^{1000} = (2^6)^{166} \cdot 2^4 \equiv (-1)^{166} \cdot 16 \ (\text{mod } 13)$$

$$\equiv 1 \cdot 3 \ (\text{mod } 13) \equiv 3 \ (\text{mod } 13).$$

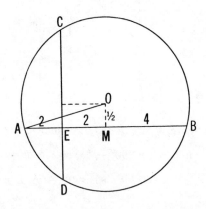

32. (B) Since point E divides every chord through E into two segments whose product is constant,

$$CE \cdot ED = AE \cdot EB \quad \text{or} \quad CE \cdot 3 = 2 \cdot 6, \quad \text{so} \quad CE = 4.$$

Thus chords AB and CD have lengths 8 and 7, respectively

Now the center O of the circle lies at the intersection of the perpendicular bisectors of chords CD and AB which is $\frac{1}{2}$ unit above and 4 units to the right of point A. The radius OA is the hypotenuse of a right triangle with legs $AM = 4$ and $OM = \frac{1}{2}$;

$$OA^2 = AM^2 + OM^2 = 4^2 + (\tfrac{1}{2})^2 = 65/4.$$

Thus the length of the diameter is $2OA = 2\sqrt{65/4} = \sqrt{65}$.

33. (C) Let the units, tens, and hundreds digits of the required number be denoted by U, T, and H respectively. Then the quotient to be minimized is

$$\frac{U + 10T + 100H}{U + T + H} = \frac{U + T + H + 9T + 99H}{U + T + H}$$

$$= 1 + \frac{9(T + 11H)}{U + T + H}.$$

If $U < T$, the quotient can be made smaller by interchanging T and U. Hence for the minimum, we have $U > T$. Similarly $U > H$, from which we conclude that neither T nor H is equal to 9. Hence our quotient is least when $U = 9$ regardless of the values of T and H. To minimize the quotient, we minimize the fraction

$$\frac{9(T + 11H)}{9 + T + H};$$

or, equivalently, one-ninth of it:

$$\frac{T + 11H}{T + H + 9} = 1 + \frac{10H - 9}{T + H + 9}.$$

This is least when T is largest, and since $T \neq 9$, we take T to be 8. Now $(10H - 9)/(H + 17)$ is smallest when H is smallest, i.e. when $H = 1$. Hence the required number is 189 and the value of the required minimum quotient is $189/(1 + 8 + 9) = 189/18 = 10.5$.

34. (A) Let T, D and H represent the ages of Tom, Dick and Harry, respectively. We are told that

$$3D + T = 2H \quad \text{and} \quad 2H^3 = 3D^3 + T^3$$

or, equivalently, that

$$2(H - D) = D + T \quad \text{and} \quad 2(H^3 - D^3) = D^3 + T^3.$$

The last equation, in factored form, is

$$2(H - D)(H^2 + DH + D^2) = (D + T)(D^2 - DT + T^2).$$

Dividing both sides by the equal numbers $2(H - D) = D + T$ yields

$$H^2 + DH + D^2 = D^2 - DT + T^2$$

which is equivalent to

$$T^2 - H^2 = D(H + T) \quad \text{or} \quad (T + H)(T - H) = D(T + H),$$

so that $D = T - H$, or $T = D + H$. By the first equation, $T = 2H - 3D$, so $H = 4D$. Since H, D are relatively prime, $D = 1$ and $H = 4$. Then $T = D + H = 1 + 4 = 5$ and

$$T^2 + D^2 + H^2 = 5^2 + 1^2 + 4^2 = 25 + 1 + 16 = 42.$$

35. (D) We first show that the triangle has to travel around the square 3 times before its vertices are again in their initial positions. The triangle first rotates about midpoint B of the square making $\frac{1}{3}$ of a revolution, then about corner X making $\frac{1}{12}$ of a revolution, and so on along each side of the square. When, after 8 moves, it reappears on side AX, it has made $4 \cdot \frac{1}{3} + 4 \cdot \frac{1}{12} = \frac{5}{3}$ of one revolution; but in order to be in its original position, with P above side AB, it must make an integer number of revolutions, and this happens after 3 such cycles, that is, after $8 \cdot 3 = 24$ moves. In $\frac{24}{3} = 8$ moves, the rotation is about P so P traverses no path; while in the remaining 16 moves, P traverses $\frac{1}{3}$ of the circumference of a circle of radius $AP = 2$ in 8 of the moves, $\frac{1}{12}$ of this circumference in the other 8 moves. Hence the total length of P's path is

$$\left(\tfrac{8}{3} + \tfrac{8}{12}\right)4\pi = \tfrac{40}{3}\pi \text{ inches.}$$

Comment: In the original statement of this problem, the last clause of the second sentence read

"until P returns to its original position".

Many alert solvers noticed that P's path goes through P's original position during the 9th move of the triangle (see figure); that is, long before the entire triangle returns to its original position. They found the length of this shorter path to be $16\pi/3$, failed to find this result among the choices offered, and raised their objections. Evidently the poser of the problem considered positions of *P* *after* each move, rather than *during* a move, and was not aware of other interpretations. The objections to the problem were justified, and the members of the Committee on High School Contests are grateful to the many people who called the ambiguity to their attention.

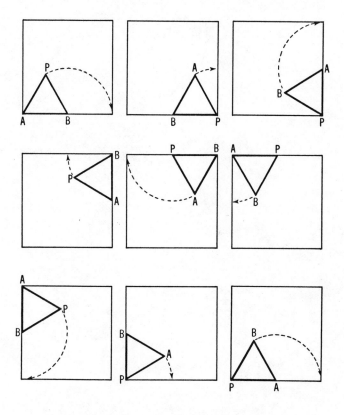

IV

Classification of Problems

To classify these problems is not a simple task; their content is so varied and their solution-possibilities so diverse that it is difficult to pigeonhole them into a few categories. Moreover, no matter which headings are selected, there are borderline cases that need cross-indexing. Nevertheless, the following may be helpful to the reader who wishes to select a particular category of problems.

The number preceding the semicolon refers to the last two digits of the examination year, and the numbers following the semicolon refer to the problems in that examination. For example, 69; 13 means Problem 13 in the 1969 examination.

Algebra

Binomial expansion	69; 16 71; 13, 24
Equations	
Cubic	67; 35 70; 11 72; 22
Diophantine	66; 26 67; 15, 24 68; 19 69; 19 71; 25 72; 33
Fractional	66; 5, 33 69; 1
Linear	66; 26 67; 19 68; 9, 34
Parametric	69; 29 70; 3
Polynomial	66; 10, 30 71; 22 72; 22
Quadratic	66; 3, 17, 23 67; 17 68; 9, 13, 14 69; 5, 7 70; 14 71; 20 72; 16
Systems of	66; 17, 22, 26 68; 14, 34 69; 35 70; 3, 35 71; 19 72; 9, 11, 34

Arithmetic

Geometry